ENVIRONMENT AND SOCIETY

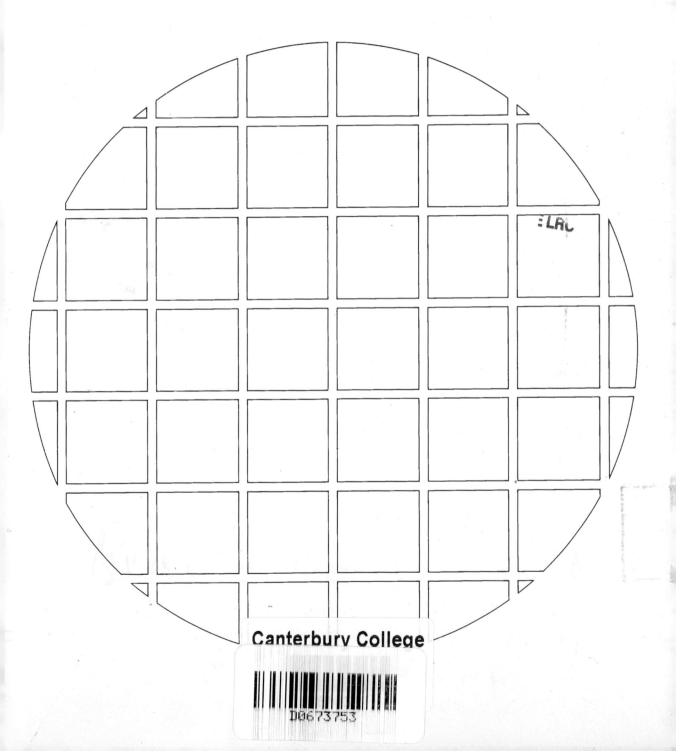

This book is the first in a series published by Hodder & Stoughton
in association with The Open University.

Environment and Society
edited by Philip Sarre and Alan Reddish

Environment, Population and Development
edited by Philip Sarre and John Blunden

Energy, Resources and Environment
edited by John Blunden and Alan Reddish

Global Environmental Issues
edited by Roger Blackmore and Alan Reddish

The final form of the text is the joint responsibility of chapter authors,
book editors and course team commentators.

ENVIRONMENT AND SOCIETY

EDITED BY
PHILIP SARRE AND ALAN REDDISH
FOR AN OPEN UNIVERSITY COURSE TEAM

Hodder & Stoughton
A MEMBER OF THE HODDER HEADLINE GROUP

IN ASSOCIATION WITH

The Open
University

A catalogue record for this title is available from The British Library

ISBN 0–340–66355–3

First published in the United Kingdom 1990. Second edition 1996.

Impression number	10 9 8 7 6
Year	2000

Edited and designed by The Open University.

Index compiled by Sue Robertson

Typeset by Wearset, Boldon, Tyne & Wear.

Printed in Spain for Hodder & Stoughton Educational, a division of Hodder Headline Plc., 338 Euston Road, London NW1 3BH, by Graphycems, Spain.

This text forms part of an Open University second level course, U206 *Environment*. If you would like a copy of *Studying with the Open University,* please write to the Central Enquiry Service, PO Box 200, The Open University, Walton Hall, Milton Keynes, MK7 6YZ.

Contents

Introduction

This book is the first in a series of four which present an interdisciplinary and integrated explanation of environmental issues. Throughout it is stressed that environmental issues are complex and that to understand them one must combine scientific evidence and theory, analysis of social processes, knowledge of technological possibilities and awareness of underlying value positions. The series considers a range of environmental issues, some local, many transnational and some global. In doing so it aims to widen readers' *awareness* of environmental issues, to deepen their ability to *analyse* them and to equip them to *evaluate* policies to influence them. It stresses that solutions to particular problems should be complementary with, and ideally contributory to, the solution of international and global problems, including global warming.

This book argues that we cannot satisfactorily understand any particular environmental issue that happens to be in the news, let alone hope for a solution to global problems, without a basic understanding of the wider contexts within which events occur. Three aspects of the wider contexts are explored here and each allows us to consider the development of current situations over time. The three aspects are: (1) ecological analysis of life on earth; (2) the impacts of human society on nature; and (3) the way attitudes and values influence those impacts.

Ecological analysis can help us understand both how natural processes work in the absence of human intervention and how nature will respond to particular forms of intervention. Part of the treatment of ecological issues deals with the relationships between the physical environment and living things at the present time. But we also ask how the physical environment came to be as it is, and the answer is a dramatic one: vital parts of today's environment for life result from the activity of living things in the past. Living things and their environments are mutually interdependent.

Examination of the impacts of human society on environments shows that this relationship is increasingly one of mutual dependence. The earliest human societies lived by hunting and gathering and had only modest effects on natural processes. However, use of fire, management of grazing animals and clearance for agriculture had very significant effects on animals and vegetation even before the dramatic increase in impacts brought about by industrialisation and use of fossil fuels. Today, human society has enormous power to alter its environment, and hence can alter its own life support system, for better or worse.

Many contemporary attitudes and practices were developed when society was dependent upon the local environment but had small impacts upon it. The biblical injunction to 'be fruitful and multiply' and later economic systems which rewarded the exploitation and domination of nature were dependent on the assumption that exploitation would not destroy the environments people depended upon. Rising population, more powerful technology and better scientific understanding have combined to show that the assumption cannot hold. As a result, there has been an upsurge in environmentalist attitudes, but these are often poorly thought out and rarely related to clear policy proposals. Conversely, while the policies of governments and corporations have reacted to environmentalist views, this is often a cosmetic change which does little to change their underlying attitudes or goals.

The authors of this book believe that putting environmental issues into their ecological, socio-technical and value contexts will help to develop a more integrated understanding, but this is not to say it will resolve all difficulties. Scientists remain ignorant of some basic information (including how many species of animals there are and what happens to the carbon dioxide released into the atmosphere); they debate among themselves about how to explain or interpret many events and processes; such debates are even more prevalent among social scientists (for example, historians cannot agree how important were social breakdowns, climatic changes and technical failures in bringing about the change of Tunisia from wheatlands in Roman times to present-day desert); moral philosophy cannot resort to evidence or logic to resolve arguments about what ought to be done. In the face of such uncertainty, the series sets out to provide the most authoritative information about what is known and what debates exist and to involve readers in making their own interpretations and defining their own value positions. Environmental problems are too important to be left to academics or politicians: every citizen has a duty to consider their own lifestyle and whether they should become actively involved in promoting or resisting particular policies. So the series aims to encourage well-informed and critical reflection, leading up to reflection about the future of human society and the planet.

This book outlines the ecological, socio-technical and value contexts of environmental issues, and later books in this series look in more depth at particular areas of concern. The second book focuses on the growth of world population and the paradoxical role of economic development as the only acceptable way of stabilising population, but a process which multiplies society's impacts on environment. This paradox is explored at length as it applies to agriculture and urbanisation. The third book examines advanced technology, both in relation to the impact of current energy and mineral use and as a source of alternative systems with fewer damaging impacts. The issue of environmental politics is also raised, introducing a theme developed in the last book in the series. Here the focus is on global issues, both from the point of view of the scientific problems in understanding oceanic and atmospheric change, including ozone depletion and climatic change, and from the point of view of the political initiatives needed if society is to regulate use of ocean resources and impacts of pollutants on the air. In conclusion, it is stressed that human society will have to define and implement the concept of sustainable development if the future is not to be 'nasty, brutish and short'!

However, most of us are daunted by the complexities of the global future and preoccupied with the local and the short term. You may be unconvinced about the need to combine such a wide range of perspectives to come to grips with environmental issues. So this book starts with a commonsense look at one area in Britain, the county of Cumbria, and argues that commonsense views of environmental problems as seen by the media actually beg many questions about how current situations have come about, how natural processes do or might work, how the unintended results of everyday activities combine with a range of management policies and above all how individuals, interest groups and communities ought to react. In trying to provide more convincing answers to such questions, the chapter shows that they require much more knowledge of how natural processes have worked in the past and do so now; of how society works and applies technology to use environmental resources and manage impacts upon it; and of how consideration of value positions can clarify practical choices.

Meanings of the term 'environment'

The term environment is used in a variety of ways in different contexts and will be extensively discussed in later chapters. But before you start to study you will need a basic understanding of the ways in which the term is used in this book.

The term originally denoted 'surroundings', but it has increasingly been applied not to surroundings which just happen to be passively there but to objects, individuals or processes which interact in some way with whatever is being surrounded.

The object being surrounded can be defined in a host of different ways, from a single individual to all living things. At those extremes, the environment is easily defined.

- An individual's environment is its life support system – the objects, individuals and processes which provide water, food, air and shelter. Clearly, a human being's environment is much more extensive than that of an insect or plant, in that the human depend on other people for companionship and in that human food and shelter depends on a world system of agriculture, industry and trade and on structures built in the past.

- The environment of all living things is composed of the physical objects and processes of the earth, seas and atmosphere, including the incoming solar radiation which is vital to life.

Between these two extremes, most usages of the term 'environment' explicitly or implicitly define it in human terms. Because human beings now exist in societies, most of which use technology and complex economic and political arrangements to produce and distribute food and water and shelter, it usually makes more sense to define *society's* environment rather than that of the human *species*:

- Society's environment is made up of natural processes (e.g. of geology and climate) and physical and biological processes themselves altered by society (e.g. through the cultivation of domesticated plants and animals).

Because the concept of environment now stresses vital connections between object and environment, it is to some extent arbitrary where the boundary is drawn between society and environment. Indeed, this book argues that the meaning of the term environment has evolved away from one which assumed that human beings were separate from nature, though surrounded by it, towards one which assumes that human beings are a part of nature and that human society is dependent on physical and biological processes for its continued existence.

1   Introduction

The aim of this introductory chapter is to start with a commonsense view of environmental issues, and to go on to develop an organised agenda which will enable you to study the relationship between environment and society with greater understanding. I have chosen the English county of Cumbria as an illustrative case study, partly because it presents a wide range of environmental processes and issues, but also because the area, especially the Lake District, has been identified by leading historians as having long been 'a forcing house for new ideas about the proper relationship between man, property, morality and environment' (Marshall & Walton, 1981, p. 219). Although the idea of a forcing house is a stimulating one, the use of the term 'man' is unfortunate, both for its implicit sexism and because it suggests that humanity exists as an abstraction rather than in highly organised societies. Nevertheless, Cumbria illustrates a range of relationships between an advanced technological society and its environment, and shows that these relationships have been influenced by different levels of understanding about how environmental processes work and by different values about the way society should relate to environment.

Setting the Lake District in the wider context of Cumbria should widen your *awareness* of the connections and contradictions between different forms of exploitation and conservation of environments. The chapter introduces a range of *analyses* – geological, historical, economic, and political – which can begin to give a better understanding of past and present environmental processes and which are developed in later chapters. Understanding environmental processes and issues demands objective and rational analysis, but policy-making is also influenced by vested interests and values. You will therefore be encouraged to evaluate the pros and cons of actual or proposed policy and to consider what *action* you might take.

The focus of the chapter is on contemporary issues, but it is stressed that to gain a deep understanding of these issues they have to be put into appropriate time contexts. The present landscape is influenced by geology (developed over hundreds of million years), glaciation (extending over hundreds of thousand years), soil formation and vegetation change (ten thousand years) and human impacts (several thousand years, but accelerating over the last century).

The argument of the chapter is developed through three sections, each of which deals with all four key questions and identifies the need for a deeper level of analysis in the next. Section 2 considers commonsense views of environmental issues in Cumbria, as embodied in a variety of newspaper and magazine reports, and begins to piece together a picture of the area and the varied demands being made of it. Section 3 steps back to analyse how the current situation came about, looking at natural science explanations of landforms and resources and also at historical, economic and political analyses of settlement and land use. In the process, several strands of relationship between environment and society are identified. Section 4

As you read this chapter, look out for answers to the following key questions.
● What kind of environmental issues exist in Cumbria?
● How did they come about?
● What policies are being used to combat the problems, and how successful are they?
● What more would you need to know to understand these issues better?

focuses on the most hotly debated environmental issue in Cumbria: the hazards of radioactive pollution from the Sellafield complex of nuclear reactors, storage and reprocessing plants. The significance of Sellafield extends far beyond Cumbria because it handles wastes from every UK nuclear power station, imports spent fuel from as far away as Japan, and has contributed plutonium for the manufacture of nuclear weapons in the UK and for export to the USA. It is therefore a telling reminder that the causes of environmental problems can involve military, political and economic motives, so that solving such problems requires much more than scientific understanding or technological capability on their own.

The chapter considers several subjects as a way of introducing the agenda for later chapters, and this has implications for the way you should study it. Try not to become preoccupied with detail but look for the overall significance of the range of issues covered. 'Activities' are included as a way of encouraging you towards the strategic questions, but in an area as contentious as environmental problems and policies you should be setting out to build your own understanding and should question views put forward by all authors, including this one.

2   Commonsense views of the Cumbrian environment

2.1   Introduction

From a British viewpoint, and that of many readers of English literature throughout the world, Cumbria means the Lake District – the area of England's highest mountains and largest lakes, which became the second British **National Park** in 1951 (Plate 1; Figure 1.1). Closer acquaintance brings the realisation that the county also contains major road and rail links to Scotland, old market towns like Carlisle and Kendal, industrial towns like Barrow-in-Furness and Workington, the Solway plain, Eden Valley and part of the Yorkshire Dales National Park. From a European or Irish perspective, probably the most striking feature of the county is the Sellafield works, which has for decades been the biggest source of radioactive discharges in Europe, with detectable effects not only in the Irish Sea but also in the North Sea. So, some people focus on natural aspects of the Cumbrian environment and others on the artificial: some come to admire and others to condemn. The purpose of this section is to use extracts from newspapers and magazines to widen your awareness of environmental issues in Cumbria, to illustrate some basic principles and to pose questions for analysis later in the chapter and the series.

Activity 1

Read the extract by the National Park Officer in Figure 1.2, then answer the following questions.

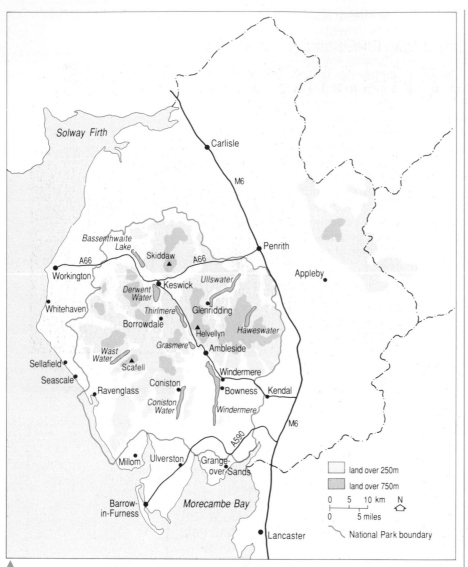

Figure 1.1   Map of Cumbria.

Q   On what grounds was the Lake District National Park nominated to UNESCO (United Nations Educational, Scientific and Cultural Organisation) as a World Heritage Site?

A   The Lake District was nominated to UNESCO as worthy of being both a natural and a cultural World Heritage Site.

Q   What was the outcome?

A   It was not accepted on the grounds that controls were not adequate to safeguard it, though the possibility of designation as a cultural site was left open. What many regard as England's foremost *natural* landscape was judged by the international body to be remarkable as an *inhabited* landscape, and to be inadequately protected in spite of the deployment of all the governmental and pressure group powers currently available.

Lake District receives world-wide attention...

THE INTERNATIONAL STAGE

THE LAKE DISTRICT National Park has received much international attention and praise in recent months. Last autumn representatives from 43 countries attended a symposium in Grange-over-Sands at which the problems facing protected landscapes were discussed. National Park and Government representatives from as far away as South America, Nepal, Japan and Australia

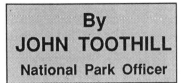

By
JOHN TOOTHILL
National Park Officer

came together to seek help and advice from one another and to see the Lake District.

The difficulties are very varied: from the loss of firewood in the Annapurna National Park through heavy tourism to education needs in Africa where sometimes the literacy rate is only 20%. Yet all these people were determined to solve these problems without harming the landscapes which had been specially designated as outstanding.

"Change must be guided so that it does not destroy"

Apart from sharing experiences the delegates also joined together to issue a Lake District Declaration which calls upon government bodies to help in the work of protection - not to create museum pieces but to ensure harmonious development. A key passage in the declaration reads:

"These inhabited landscapes are in delicate equilibrium; they cannot be allowed to stagnate or fossilise. But change must be guided

National Park field staff keep in close touch with the farming community.

so that it does not destroy but will indeed increase their inherent values. This means for each protected area a clear definition of objectives, to which land use policies within it should conform. It means also a style of management that is sensitive to ecological and social conditions. This will be possible by building upon spiritual and emotional links to the land and by the operation of flexible systems of graded incentives and controls."

In December the nomination of the Lake District as a World Heritage Site was discussed in Paris by the World Heritage Committee, a body appointed by UNESCO to give recognition to those places in the world which, "because of their exceptional characteristics and qualities, should be formally listed as being of outstanding universal value."

The Lake District had been proposed as a site by the UK Government on both the grounds which merit recognition: the natural heritage and the cultural. This was unusual - normally a site only qualifies under one heading (e.g. Hadrian's Wall is a cultural site; the Giant's Causeway is a natural site), and the Committee decided to defer the Lake District nomination because the controls in this country over agricultural, forestry and military operations do not match up to their requirements. But the Lake District is to be reconsidered for designation as a cultural site.

So the Lake District has received world-wide attention, and I have been impressed with the knowledge of our affairs shown by overseas workers in landscape protection. They know of our erosion problems and the work of the Upland Management teams; they know of the second home difficulties and our attempts to alleviate it; and they look to the Lake District as a place where answers are being sought to problems which they might well face in the future.

LAKE DISTRICT NATIONAL PARK VISITOR CENTRE
BROCKHOLE WINDERMERE (096 62) 6601

The National Park's own visitor centre overlooking Windermere is Lakeland's premier attraction for locals and holidaymakers of all ages. Ask for free "What's On at Brockhole" brochure from any information centre.

Open daily Easter to 6th November 1988
Season, weekly and family tickets
Adults £1.30p: 5-17 65p: Under 5 free.

Brockhole is on the A591 between Windermere and Ambleside
(More details see page 22)

ON OTHER PAGES

£1 million Footpaths Bill	3
The Big Litter Sweep	5
Holidaymaker Facts	5
Land of Sunsets	7
Solution in the Air	7
Saving our Bridle Roads	17
Wardens' gratitude	17
Seats and Memories	17
War on Grey Squirrels	19
This new World	19
A Vote-winning View	24
Arts from the Lakes	24

Including eight page pull-out carrying a comprehensive list of the important events taking place in the region from April to October.

Figure 1.2   Extract from The Lake District Guardian, 1988.

Q   What claim did the author make after juxtaposing the positive support received from international delegates with the negative response from UNESCO?

A   The claim made at the end of the piece is that the Lake District is in the lead in developing policies not only for landscape **conservation** and tourist use, but also to meet the needs of a substantial resident population. The strength of the claim is somewhat weakened by the simultaneous attempt to use these events to provide authoritative international backing for another view – that more needs to be done by government bodies to create a structure of incentives and controls to conserve the area, as they have done in other parts of the UK and in other countries.

The implications for this course are clear: if even England's foremost upland area is not acceptable as a natural site, the whole of England (and arguably of Scotland, Wales and Northern Ireland) has to be thought of as an area where natural environmental influences interact with human activities to a greater or lesser extent. Later chapters will go on to show that the same is true of much of the world and that even many areas that we think of as wilderness are subject to significant and growing impacts. To understand environments we need to combine the insights of natural science, technology, social science and the humanities.

2.2   Agriculture

In the Lake District, as elsewhere, one of the most extensive ways in which society draws upon, and changes, the natural physical environment is through agriculture. Much of the attraction of the area stems from the contrast between small green fields and open uplands, the use of local stone and slate in dry stone walls and farm buildings, and the sheep and cattle that move between upland and lowland pastures (Plate 2). This pastoral landscape has resulted from the elimination of the natural vegetation cover, which would be of broadleaved woodland up to all but the highest or most exposed summits.

> *Activity 2*
>
> Read the extract about the fell farmer (Figure 1.3), then answer the following questions.

Q   What are the problems of farming in the Lake District?

A   The natural problems identified are that land is 'rocky and poor' and that the upland situation also gives rise to a cold, wet and windy climate and steep slopes. Although tourists are a source of extra income, they also bring damage and disruption.

Q   What organisations provide farmers with assistance?

A   The farmer is assisted by the National Trust (a private charity considered later in this section) and the national park authority, a part of the system of local government, formally entitled the Lake District Special Planning Board.

Q   What do you think are their goals?

A   Their goals focus on the preservation of the existing landscape (from details such as stone walls, slate roofs and wooden gates to the overall look of sheep-grazed fells) and the way of life of the farming community.

In recent decades, hill farmers have also been supported by subsidy payments on sheep and cattle, at first from the UK government and more recently from the European Community's Common Agricultural Policy

Fell farmer trusts his homing instincts
by DAVID WINPENNY

FOR 18 YEARS Daniel Birkett has been a tenant of the National Trust in the Lake District. Of the 700 acres of Yewtree Farm less than a sixth can be classed as good land: the rest is enclosed fell, which provides grazing for about 200 sheep.

Dan doesn't deny that life here can be tough for the farmer. "Land round here is very rocky and very poor" he admits. "The sheep we sell are always at the lowest end of the scale, compared with those from better farms – and farmers on better land get just the same subsidies." He's often told that, although he does very well for the area, he would be better off if he moved.

So why does he stay? "Because I was born here. It's a bit like being a Herdwick sheep on the mountain. It's your homing instinct that keeps you here. And, of course, I like the scenery."

But beautiful scenery can bring its own problems. Dan's land, spread in small parcels around Coniston, reaches as far as Tarn Hows, one of the most visited and photographed of the Lake District's beauty spots. The visitors can be a nuisance, although, he says, only at certain times of the year. "At Easter it's as if they've just been let out of cages," he says. "They come up with their dogs and chase your sheep. From then until the beginning of summer is the worst period. Once they've got themselves broken in, they're not so bad again."

The vast majority of the visitors keep to the footpaths, of which the farm has many, and of those who wander off nearly all will respond to a few quiet words. And Dan has learned to live with the occasional wall being damaged or fence climbed. Sometimes he will come across those who say "This is a national park; we can sit in your mowing grass and have a picnic." Dan deals with them by explaining that the grass is his living.

Repairing damaged walls doesn't bring in any profit –"We always say it's a halfpenny-an-hour job" – but as Dan points out, "It has to be done to keep the Lake District right. Once you start saying 'That wall's down – I'll stick a bit of tin in here', you're finished."

Being a National Trust tenant means that Dan has to be careful with his buildings, too. He was busy altering a cubicle house for his beef suckler cows, and was having to ensure that it was done to the Trust's specifications. The Trust helps with grants for the work, but he still finds that, to keep costs down, he has to do a lot of the work himself.

He worries, too, about increases in rents. Every three years there is an increase which, he says, is too often. "Farming in the Lake District doesn't alter every three years. If they came every seven or nine years it would be enough. The price of sheep doesn't alter every year. We're in an area where a bad summer or winter can affect you for two or three years." He thinks that the Trust should look further ahead. He has seen other farms use too much fertiliser on the land, or introduce wrong crosses of sheep into the area. Some try to keep too many sheep, causing friction with neighbours – and all to try to generate extra cash in the short term, while damaging both the landscape and the harmony of local communities.

One way to supplement income is to take tourists – and Yewtree Farm is open part of the year for bed and breakfast. This is largely the province of Jean, Dan's wife (though he is quick to point out that she also does every bit as much work on the farm as he does). They can take about six guests at a time, and have made many friends in this way.

Dan doubts whether there will be an increase in tourist numbers, but thinks

that, with increased leisure, and the ease with which visitors can reach the area, seasons will be longer, with short winter breaks particularly important.

As for the national park authority, Dan thinks it is doing a good job. Although he complains about the amount of paperwork the upland management scheme can create, he says "I think the upland management, especially in this area, has done an excellent job. Yewtree Farm has between 40 and 50 gates, for example, all to maintain, and half of them used by the public; if there wasn't a scheme to help with them it would be pretty hard work keeping them in good order."

He agrees with the restraints on farm buildings which development control brings. "In certain areas there shouldn't be these big nasty-looking buildings put up and old ones left to go rotten, with broken slates tumbling down and rotting away. A lot of these old buildings should be maintained for future generations."

A traditional Lake District spinning gallery is a feature of the yard at Yewtree Farm, and visitors show a great deal of interest in it. Dan doesn't mind people photographing or drawing it. "It's all part of life, and that's the way it should be done." Magician Paul Daniels filmed part of his show under the gallery recently, drawing a good crowd to watch.

Like all upland farmers Dan Birkett struggles to make a decent living. He has no illusions about it. "I wouldn't advise anyone to farm in the Lake District, because it's damned hard work" he says. But he remains where he was born, keeping his farm up to the standards which not just the National Trust or the national park authority but he personally feels should be maintained "to keep the Lake District right".

Figure 1.3   Extract from National Parks Today 16, Winter 1987, Countryside Commission.

and in particular its scheme for 'less favoured areas' – i.e. those suffering from environmental disadvantages. Unfortunately, the National Trust and national park authority schemes fall far short of the level needed to insure farmers against loss of subsidies as a result of the EC reacting to the problem of surplus production.

2.3   Mining and quarrying

Another use of the land which has had major effects over several centuries is mining and quarrying. The Lake District contains ore deposits of several metals, including copper and lead, as well as first class stone and slate. The Furness area contains deposits of haematite iron-ore which were the largest known in the world for a period in the nineteenth century. The west Cumberland coast has many coal mines, though most have proved uneconomic and have closed in recent decades. Mining and quarrying have had major direct effects on the landscape – quarries, spoil heaps, industrial buildings, stream and lake pollution. They have also had indirect effects: felling of trees to provide charcoal for smelting and urbanisation, prompting rail and road building, and the use of lakes, rivers and streams for water supply. Within the National Park many of the worst effects have been softened by time, and mining towns like Coniston and Glenridding have become tourist centres. But in other parts of Cumbria derelict land and stagnating industrial towns form a stark contrast to the idyllic landscapes of the Park.

2.4   Housing and tourism

The decline of mining and quarrying employment has created a supply of surplus buildings and eroded the incomes of local residents. This, together with sales of surplus farm buildings as farms amalgamate to produce larger units, has encouraged the rapidly growing demand for retirement, second or holiday homes in the Lake District. More than half the parishes in the South Lakeland sector of the National Park have a fifth or more of their housing as second or holiday homes. This may bring in income in the summer months, but may damage the village community, reduce demand for local shops and schools, and make it impossible for local people to find housing they can afford.

Activity 3

Read the extract on Chapel Stile (Figure 1.4) and answer the following questions.

Q   What policies have the Lake District Special Planning Board tried or proposed to provide housing for local people?

A   Originally, there was an attempt to increase the supply of dwellings. More recently there have been attempts to use administrative rules to prevent sales to outsiders.

Q   What is the root cause of the problem?

Death of a village
Chapel Stile cited as typical ghost village

Anyone who believes that the second home and holiday home problem has not damaged Lakeland should take a drive or a walk up Great Langdale on a dark night, it was stated this week.

They should carefully count the number of houses they can see outlined against the night sky – and then go back and count how many have lights at the windows.

For Great Langdale, and the hamlet of Chapel Stile in particular, are a prime example of a Lakeland "ghost village", said Coun. Tom Shelton, chairman of the Lake District Planning Board's Development Control Committee.

There are a vast number of second homes in the valley and it has spoiled the environment because young, local people have been forced to move away and it has resulted in the closure of shops, loss of buses and so on," he told a public meeting in Grasmere.

Coun. Shelton had been asked by Methodist Minister, the Rev. Norman Pickering, if the Board was doing anything to defuse the growing tension between local people and those who had second homes in Lakeland.

And he said: "It is a question that has been asked quite a few times over a number of years because it's true there is a conflict of interests here, and we are trying to find some solution to provide homes for local people who cannot compete in the open market with would-be second home owners.

"We are trying to impose local occupancy conditions or Section 52 agreements onto all new houses being built to prevent them being sold outside the area."

Coun. Shelton said it was already too late to save some communities, and Chapel Stile in Langdale was an obvious example.

"Walk up Langdale on a winter night and you will hardly see a light at any window," he suggested. "It is a ghost village which we will just have to live with."

Principal Administration Officer, Mr John Toothill, pointed out that the Board had tried to help Chapel Stile 11 years ago by allowing 27 "cheap" modern homes to be built on a former quarry site to encourage locals to buy – but the plan had not worked.

Of the 27 homes built, only one was in permanent occupation and the rest were second or holiday homes.

"It was after that experiment that we realised that simply allowing more homes to be built would not work because locals did not buy them anyway. It is an experiment we are not likely to repeat."

▲   *Figure 1.4   Extract from* The Westmorland Gazette, *6 March 1981.*

A   The root cause is increasing demand for second or retirement homes from affluent urbanites who can afford to outbid most locals for any property on the open market.

One response to this situation, which takes some of the pressure off the existing housing while meeting growing demand from tourism, is the construction of timeshare accommodation. The first major timeshare development was the Langdale estate which, as their advertisement states (Figure 1.5), won a design award. It gained local support because it improved a derelict site and provided numerous jobs. However, it also included extensive indoor recreation facilities and has begun to acquire cottages, so some negative impacts exist. Later proposals have begun to threaten large-scale development in much more conspicuous sites. For example, there was a proposal in 1988 for a development which would have accommodated over a thousand people on the shores of Windermere, again with a large indoor leisure complex. Some of the timeshare proposals escape planning control because they involve upgrading of former caravan sites, where existing planning permissions allow conversion to chalets. In these circumstances the planners can do little more than bargain with developers to modify the worst impacts on the landscape.

Langdale – Europe's most sought-after time ownership. Discover why.

Winner of the Civic Trust Award for its contribution to the quality of the environment, the exclusive Langdale private estate offers unrivalled amenities in 35 wooded acres of privileged seclusion in the Lake District National Park. Luxurious Scandinavian lodges. Sauna, whirlpool bath, luxury kitchen, B&O hi-fi, satellite TV.

Langdale time ownership includes membership of the £3 million Pillar Club, an all weather leisure oasis with its tropical pool, squash, hydro-spa, gym, beauty salon. Real-ale pub. Gourmet restaurant. Beatrix Potter and Wordsworth country on your doorstep.

Langdale helps you get the most from the Lake District you love. A single payment now and it's yours for life, to enjoy or exchange for holidays in 1,400 resorts around the world.

Phone or post the coupon. Your 20-page colour brochure will arrive without obligation. Better still, come and discover for yourself why Langdale is the most sought-after timeshare in Europe.

☎ **Langdale (09667) 391 (24 hours).**

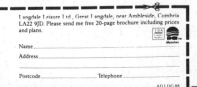

LANGDALE
SHARE IN THE BEAUTY OF THE LAKES

▲   *Figure 1.5   Advertisement for timeshare ownership.*

2.5   The National Trust and preservation

The difficulties of the Lake District Special Planning Board in managing a park which is largely privately owned, and where the economic opportunities involved in tourism threaten the very qualities which attract tourists in the first place, confirm the foresight of the founders of the National Trust. For a period in the late nineteenth century, Canon Rawnsley, Octavia Hill, Robert Hunter and others had been involved in a variety of pressure groups resisting development in the Lake District and elsewhere. They then came to the conclusion that the only way to be certain of conserving landscape was to own it, and in 1893 established the National Trust to hold land and buildings in perpetuity. A century later, the National Trust owns a quarter of the National Park, which is their most concentrated area of interest.

Activity 4

Read the National Trust leaflet reproduced as Figure 1.6 and answer the following question.

THE NATIONAL TRUST AT WORK IN THE LAKE DISTRICT

The National Trust was founded in 1895 to help preserve, for everybody and for all time, the beauty and character of our national heritage.

The Trust today protects 500,000 acres of land in England, Wales and Northern Ireland, 460 miles of Britain's unspoilt coastline and over 200 historic houses open to visitors.

As part of this, the Trust protects more than a quarter (140,000 acres) of the Lake District National Park, including almost all the central fell area and the major valley heads, 6 of the main lakes and 7,000 acres of woodland.

Farms and Cottages
The protection of the Lake District and its beauty is inseparable from the farming system which has evolved over the centuries, and the support of the Trust is essential to the preservation of the fell farms.

The Trust has spent £850,000 on its farms over the last 3 years, and £700,000 on its cottages.

It looks after 80 working farms, let to local people, often as their first step on the farming ladder, and over 250 houses and cottages, again nearly all let to local people, making the Trust the second largest provider of rented accommodation to local people in Cumbria after the local councils.

Footpaths and Erosion
Millions of us look to the mountains to escape from the turmoil of our everyday lives. But how much pleasure or freedom is there in sweating up an eroded river of stone 70ft wide created by the tens of thousands who went before?

The National Trust has already spent hundreds of thousands of pounds rebuilding the worst eroded paths,

▲ Builders re-slating a Trust barn roof at Side Farm in Patterdale

◀ Starting the repair of a massively eroded footpath

including Helm Crag, Sty Head, Grains Gill, Sour Milk Gill and many more. It is expensive work and slow – averaging 15 yards a day is hard going for a team of four. There is a long way to go yet.

Woods
The Trust has planted 150,000 trees in the Lake District over the last 3 years, costing it over £1.2 million. Each young tree costs less than £1, but, including fencing and

Work is very often slow and hard ... and wet ▼

protection against sheep and deer, each mature tree will cost nearly £20.

The ancient oakwoods of Borrowdale, the mixed woodlands of Claife, the elegant larch trees of Manesty on Derwentwater; these are just a few of the forests, woods and even individual trees maintained and renewed by Trust foresters throughout the year.

Wardens
The Trust has a team of 21 Wardens responsible for the land in its care in the Lake District. They repair footpaths, bridges and stiles, look after visitors and their safety, pick litter, even find lost cars and parents for lost children.

The National Trust was born out of the struggle to preserve the Lake District as 'a national property' over 100 years ago.

The Trust today has a unique role to play:
– the Lake District must be protected for all time; it is our responsibility to ensure that it can be appreciated and enjoyed by everybody, today and in the future.
– the Trust has a special power; nearly all of the land that it owns has been declared 'inalienable', which means that it cannot be sold or compulsorily purchased without an appeal to Parliament.

However, the Trust, as a registered charity, has to raise the funds necessary to look after this beautiful heritage. The Trust feels that the visitor and walker must be given the freedom to enjoy the magnificent countryside, and it is not Trust policy to charge for access to the fells and lakeshores in its care.

We must rely for our support and money on those who share our conviction that the Lake District must be looked after properly.

The Trust has carried out detailed surveys of the

conservation needs of the Lake District, including its farms and traditional working buildings, footpaths and walls, its nature conservation responsibilities, and the needs of the landscape and the community as a whole.

On the basis of these up-to-date surveys, the National Trust has drawn up a detailed work programme of repair, restoration and maintenance for the Lake District in the five years from 1987 to 1991. This programme will cost an estimated £12½ million.

Of this £12½ million, the Trust will be able to give, with the continuing support of its partners, £10½ million.

The Lake District Appeal is asking for your help to raise the remaining £2 million by 1991.

The Lake District is special. We need you to work with us to ensure that it stays that way.

Lord Shuttleworth, Chairman of the Appeal.

The National Trust is a registered charity. Charity Commission No. 205866.

▲   Figure 1.6   National Trust publicity leaflet.

Q How does the preservation of a landscape differ from the preservation of a historic house?

A The preservation of a historic house (not discussed in the leaflet) is largely a matter of preventing deterioration. The preservation of a landscape, as the leaflet makes clear, involves preserving the farming system as an ongoing activity. Landscape *and* houses may be subject to wear and tear from visitors, but it is easier to control this – and charge for it – in the case of a house.

In spite of problems of maintenance and funding, there is little doubt that the role of the National Trust has been crucial in preserving some of the most outstanding areas of the Park from insensitive development. In this respect it has been a much more positive influence than the two major public sector landowners in the Park, the Forestry Commission and the North West Water Authority, both of whom have at times pursued their special interests in ways which offended conservation bodies and local authorities.

2.6 Roads and traffic

Nowhere could the dilemmas of the public sector over conservation and development interests be more vividly illustrated than over traffic and road building. The growing popularity of the Lakes with car-borne tourists brings its own pressures and problems, but these are compounded by the existence of the industrial towns of the west Cumberland coast and the port of Barrow. These were areas of rapid growth a century ago, because of local deposits of coal and iron-ore, but have seen a decline in employment and depression for much of this century. One of their problems has always been poor inland communications as the rugged terrain hampers road and rail building. To improve the economic prospects of west Cumberland, it was proposed in 1969 to improve the A66 road between the M6 motorway at Penrith and Workington on the coast. The arguments against the proposal and the proposed alternative route north of the National Park

◄ *Road construction work for the A66 on the viaduct across the Greta Gorge.*

boundary are summarised by the extract from *The Sunday Times* reproduced as Figure 1.7. Since the alternative route was little longer and only slightly more expensive, it is unclear why the Secretary of State for the Environment insisted on the A66 road alignment, though it has been suggested that this may have been the price of a major industrial investment. Although the work was completed ten years ago, there is still great resentment among Lakeland enthusiasts about the routeing of a trunk road through attractions like Greta Gorge, Keswick and along the shores of Bassenthwaite Lake.

High speed carve-up of Lakeland's park
Tony Crook

JOHN TODHUNTER has farmed 160 acres alongside beautiful Bassenthwaite Lake, near Keswick, for over 40 years, ever since he helped as a lad on his father's farm. His grandfather was a farmer in the area before him and John's two sons both followed him on to the land. One is now a tenant farmer on a neighbouring plot.

"There've been a few changes since those days of mucking out the cows for my father," he says. "But nothing like what's going to happen here soon."

What's to happen soon is the improvement to the A66 through the Lake District National Park from Workington to Penrith. To John Todhunter it means the loss of one tenth of his dairy farm, the severing of a portion of his grazing land, and a lot of inconvenience – "My 40 Friesians will have to cross that 60 mile an hour road four times a day." To the rest of Britain it means incalculable and permanent damage to our premier national park.

The A66 proposals, agreed by Mr Geoffrey Rippon, Secretary of State for the Environment, just before Christmas (was that a deliberate ploy? ask the conservationists) affect 23 miles of highway inside the national park, 16 miles of which will be construction works. His decision, reached nine months after the ending of a mammoth seven-week public inquiry, is likely to remain the most controversial and regrettable for many a year.

The objectors make the following points:

1 A major highway carrying industrial traffic should not be brought through our prime national park at any cost, and that the whole policy of roads within national parks needs thinking through.

2 The vast construction and engineering work is completely alien to the close and intimate Lake District setting.

3 The dual carriageway along Bassenthwaite Lake is intolerably intrusive and would destroy its natural setting.

4 The Keswick by-pass, with its viaduct and large clover-leaf interchange, would alter the whole character of the town – and would not solve its traffic problems anyway.

5 There is a valid alternative route for industrial traffic from West Cumberland to the M6 avoiding Keswick and the national park.

6 The whole concept is inconsistent with the purpose of a national park.

The alternative route, which the public inquiry's inspector, Sir Robert Scott, accepted as "a valid alternative to the A66 as a trunk road link between West Cumberland and the M6 which industrial through traffic could use", would sweep north east from Cockermouth along the A595 then turn south east on an improved B5305 to meet the M6 just north of Penrith. The so-called Sebergham route.

TALKING to numerous people along the A66 route, I couldn't avoid the odour of political expediency. West Cumberland with 130,000 people in it, is not a large development area, but its needs are important. No objector to the A66 proposals would deny that. What does concern people is the fact that West Cumberland's past health has been based on coal and iron and things now seem to be declining. Is it possible forcefully to revive the area? If so, how long will its new lease of life last? And finally, if the revival does not last we shall be left with an indelible scar across one of the most beautiful areas of Britain.

▲   *Figure 1.7   Extract from* The Sunday Times, *7 January 1973.*

2.7   *Sellafield and radioactive pollution*

Perhaps the only issue which arouses greater controversy than major roads is that of Sellafield. This complex, which includes the Windscale reprocessing facilities and Calder Hall nuclear power station – the world's first – were started in a former Royal Ordnance Factory in 1948 precisely because of its remoteness. The initial function was to produce plutonium for the then secret atomic bomb project, so there was no question of planning inquiries at first. However, discharges of **radioactive** materials to the atmosphere and sea, on a routine basis and also as a result of accidents (notably the fire in 1957), have created considerable public alarm. The extract dealing with 'monsters of the deep' (Figure 1.8) indicates the extent to which **radiation** triggers fantastic fears and the degree to which parts of the press rely on sensationalism rather than analysis.

A more serious debate was started in 1983 by the Yorkshire Television programme *Windscale – the Nuclear Laundry* which revealed the existence of a concentration of leukaemia cases near the plant. Subsequent studies by the Black Committee and other government appointed bodies have confirmed the existence of this anomaly – and of another at the fast breeder reactor site at Dounreay in Scotland – but have not been able to account for them in terms of known exposure of local residents to radioactive materials and accepted international estimates of risk factors. The issue remains contentious – opponents of nuclear power arguing that doses and risks must in fact be higher than officially estimated and supporters arguing that there must be some other cause of the anomaly.

Section 4 of this chapter returns to Sellafield but for the moment it is more important to note that it raises a key issue for environmentalists: sometimes even the best available scientific evidence is inconclusive and policy decisions have to be taken on other bases. All too often these are dictated by inertia, prejudice, vested interest or propaganda, but they make it essential to address the question of how to make policy decisions in the face of uncertainty.

Activity 5

Start to monitor your local media (newspapers, radio and television) and build up a file of information and views on local environmental issues. Analyse them as we have done here. For example: What is being proposed (or resisted)? What sides exist and who belongs to them? What is the basis of arguments for particular proposals? Do proponents of particular views have vested interests in particular outcomes? By what procedures will the dispute be resolved? This file will help you relate the general arguments in this book to your local area.

2.8   *Summary*

Looking back over the extracts reprinted here, you should be able to see that they deal with cases in which widely differing processes are involved. Some of these are physical or biological processes, some occur within society, and most involve interactions between society and the physical environment.

'Monsters claim alarmist'

PRESS COUNCIL BACKS MP

A NEWSPAPER article about pollution of the Irish Sea was "alarmist and irresponsible," says the Press Council, upholding a complaint by Workington Labour MP Dale Campbell-Savours.

Mr. Campbell-Savours had been highly critical of a *Daily Star* story headlined "Monsters of the Deep", which referred to mutant fish in the Irish Sea. The article claimed that deformed fish had been poisoned by Sellafield radiation discharges.

Mr. Campbell-Savours told the Press Council that the newspaper irresponsibly published a misleading and alarmist article, containing inaccuracies relating to the activities of BNFL and others at Sellafield.

"I take a thoroughly objective decision on Sellafield," the MP said. "When they are wrong, we should hit them hard. But, when they are right, we should defend them.

"On this occasion they were right and I pushed the Press Council into making this decision, because such articles damage West Cumbria. We can't afford this type of damage."

The newspaper also referred to a scientist, Dr. Ian McAulay, finding abnormal levels of caesium in the Irish Sea.

On an inside story it said that Dr. McAulay, a senior lecturer in Experimental Physics at Trinity College, Dublin, would not eat fish caught in the Irish Sea.

In fact, Dr. McAulay, who has spent several years studying the effects of discharges into the sea, told the Press Council he had tested a sample of fish supplied by the *Daily Star* and told the newspaper that the results were normal.

In his opinion, the quantities of caesium which he attributed to Sellafield discharges presented no risk from their radioactive content.

He had stressed to a reporter that he would not eat the sample because of its disease, not because of its radioactivity content. His measurements of fish showed a steady decrease in radioactivity since 1980.

Mr. Campbell-Savours said the newspaper did not consider other factors which might have produced the deformities in fish. No statistical evidence was produced to link radioactivity and fish deformities and radioactivity and Sellafield. If such evidence were produced, it would be successfully disputed.

The Press Council adjudged that:
● Newspapers must be free to write about the danger from pollution of the environment, including the sea. On an issue of such importance, however, they have a particular duty to be accurate in their reports and fair and balanced in their comments.
● In this case, the *Daily Star's* article placed misleading emphasis on nuclear pollution of the Irish Sea and the effect of this on fish.
● The article exaggerated the likely effect of the radiation level found in fish in the Irish Sea, misquoted a scientist who had reported on it, and misrepresented his views.
● The article was alarmist and irresponsible, and the complaint is upheld.

▲   *Figure 1.8   Extract from* BNFL News, *November 1988.*

Physical and biological processes are important in two ways:

1   Very long-term geological processes, together with climatic changes, have created the rocks, minerals and landforms which provide the basic structure of the area, and evolution has produced living things including the human species.

2   Shorter term processes (including weather and water movement), together with plants and animals, provide continuing resources for human activities and also react, sometimes unpredictably, to changes brought about by society – for example, damage to vegetation by walkers' boots allows running water to strip off soil and turn footpaths into gullies.

Societal processes relate to environment in four ways:

1   Certain activities rely on environmental structures and processes for the **resources** they use – coal and metals for industry, climate and soils for agriculture, etc. Not all these resources are local.

2   Human activities have cumulative **impacts** on natural environments and may modify them substantially. Some impacts may be seen as positive – the 'farmed landscape' is regarded as an improvement on nature by most Lake District enthusiasts. Others may be harmful to some aspect of pre-existing environments: we have highlighted road building and radioactive pollution, but similar points could have been made about chemical works, domestic refuse, acid rain or sewage from picturesque villages like Grasmere or Hawkshead.

3   The extracts also refer to the recognised need for co-ordination and **management** of the complex range of demands and impacts that particular activities have on their environments. County and district councils, Parliament, central government and the Lake District Special Planning Board all have roles in regulating competing uses, pressure groups and a wider public interest, but may themselves conflict, and have been assessed by UNESCO as providing inadequate protection. As a result of such a plethora of organisations and so many different motives and values involved, policy- and decision-making are complex activities and crucial to environmental outcomes.

4   Much of our understanding of natural processes and the effects of societal demands and impacts rests on *scientific* investigation and theory. Unfortunately, scientific understanding of environmental processes remains limited even in areas of urgent social concern, such as pollution from Sellafield. All too often, moreover, economic and political decisions are taken in the absence of, or in defiance of, such understanding.

Finally, it is crucial to realise that an increasingly powerful role has been played by technology – from forest clearance through mining and smelting to nuclear reprocessing. Indeed, the technological capacity to intervene in natural processes has arguably developed more successfully than either the scientific ability to understand the effects of those interventions or the societal ability to regulate harmful impacts. Accordingly, the next section in this chapter will highlight technological development as a key aspect of human modification of Cumbrian environments.

3   Understanding the Cumbrian landscape

3.1   Introduction

The aim of this section is to develop a more systematic understanding of the interactions of society and environment that have created the Cumbrian landscape. Cumbria is used as an example in this chapter, but the real intention is to consider what you need to know to understand the broad context within which particular environmental issues and problems occur. In such a short account explanations cannot be exhaustive: many topics will be briefly introduced here which will be given more sustained treatment in later chapters. Once again, try not to become preoccupied with detail, but look for the broader principles.

It would be possible to set about understanding links between society and environment in an area by describing for each part of that area what environmental features existed (e.g. high or low land, steep or level, warm or cold, wet or dry) and what societal structures or activities (e.g. settlements, farms, factories, roads) were present. Such a study was carried out in Cumbria in 1989, and shows that there are associations between such features. For example, wheat fields are most likely on the flat terrain and deep soils of the Solway Plain, certain wild plants exist on upland areas of the Pennines and others on the slopes of Lakeland, and so on. Overall, a fascinating and detailed picture of different kinds of environment can be constructed. From our point of view, it has two serious disadvantages: it does nothing to show what caused the natural or social features to be as they are, and it does not tell us whether the present situation is likely to persist or whether it is in the course of **change**. People with a concern about environmental problems must be able to deal with change, to anticipate undesirable changes and/or to promote desirable change.

One way of beginning to understand environmental change is to ask how an area has developed into its present state. The study mentioned in the previous paragraph shows that the height, slope and geology of an area is a major influence on soil, plants and human activities. A very long timescale is needed to see how rocks and landforms have been constructed but much shorter time spans can show the development of human activities. The account that follows will start by looking at geology; however, a point to bear in mind when considering a long-term social history is that scientific views of geological and biological questions have been dramatically transformed within the time that society's use of Cumbria has been consolidated.

3.2   Development of the natural landscape

Introduction

If you had assembled a team of leading geologists and biologists two centuries ago and asked them to explain the Cumbrian landscape to you, they would have given you a very different picture from the one that follows. They would have agreed that the area was much as God had created it – after all, that event had happened as recently as 4004 BC. Some would have been aware that certain rocks contained fossils, which apparently were remains of plants and animals – some of which no longer existed. These had clearly been eliminated by Noah's flood, leaving room for the growth in numbers of species saved by the Ark. There might have been some debate among them as to whether God had intervened more than once in his creation, either with catastrophes such as flooding of land areas and raising up of the sea bed or by creating additional species of plants and animals. They might also have disagreed marginally about humanity's mission – was it to multiply and dominate or to act more modestly as God's steward on earth?

By the 1790s there might have been some among them who were aware of the view that, rather than looking to acts of creation or to catastrophes, geological structures and the fossil record could be explained by the operation of natural processes acting over very long periods. So the existence of thick beds of sandstones, mudstones or limestones could be the result of the laying down of thick layers of sand, mud or shells which were subsequently consolidated and then lifted, often into the highest hills or mountains. As the nineteenth century progressed, more and more scientists came to adopt this theory, and it was extended by Charles Darwin and Alfred Russell Wallace to include the theory of the evolution of successive forms of plants and animals. Initially, the majority accepted that these long-running geological and biological processes were the result of a Divine plan, and it was only in the twentieth century that most scientists came to see geology and biology as entirely natural sciences. The change of views in the nineteenth century was most vividly expressed in Cumbria in a debate over the origins of the valleys and lakes, but to establish the basis for that debate it is necessary to look first at the rocks.

Geology

To the non-geologist, then as now, the explanation of the distribution and nature of the rocks of an area seems mysterious. There are some clues: upland areas often seem to be made of harder rocks than lowlands; rivers sometimes follow bands of softer rocks; quarries often show layers of one kind of rock on top of others. But these clues are far from infallible – hard granite can be found in low areas and soft limestone on mountain tops;

rivers may cut through the hardest rock; a second quarry may show the same layers folded, slanted or even overturned. To those not initiated into its mysteries, a geological map, such as that of Cumbria reproduced as Figure 1.9, may compound the mystery at first sight. To resolve this difficulty, the account that follows will first describe some of the main rocks of the area, starting with the oldest, and suggest what conditions were like when they were formed. After that, we will need to consider how Cumbria could have experienced such a variety of circumstances.

The oldest rocks of Cumbria were laid down some 500 million years ago as muds and sands on the bed of an ancient ocean. These rocks were later deeply buried, heated, deformed and uplifted by geological processes and now are known as the Skiddaw Slates, forming the rounded hills of the northern Lake District.

The next geological event is sharply contrasting in origin and nature. An era of volcanic activity produced vast outpourings of molten rock and ash which may have approached 4 kilometres in thickness. This major episode formed the area of rocks now known as the Borrowdale Volcanic Group, which gives the rugged mountain scenery of the central Lake District. Fine volcanic ashes, consolidated by heat and pressure, became the high quality slates now widely used for ornamental and monumental purposes.

Eventually the volcanic activity died away and the sea covered the volcanic landscape. The resulting layers of sediment now form a thin deposit of lime-rich and fossil-rich mudstones, known as the Coniston Limestone Group, followed by thick sequences of muddy and gritty rocks laid down on the sea floor – the Silurian Slates. Much of this sequence was later eroded by running water, and the limestone now outcrops as a narrow band running from Millom to Shap, separating the harsh crags of the north from the more subdued landscapes around Coniston and Windermere.

At the end of this long period of sedimentation came an event which, more than any other, shaped the region. Around 400 million years ago the area was arched up into a huge ridge. This was accompanied by the injection of great masses of molten rock which crystallised under the Lake District mountains, forming a rock called granite, the upper parts of which are exposed today in the Shap and Skiddaw granites in the east and as the Eskdale and Ennerdale granites of west Cumbria. Associated with the granites are deposits of minerals, including lead and copper, which were exploited, for example, in the Coniston copper mines. However, the mountains were rapidly worn down as there was no land vegetation at that time to bind the weathered rock debris.

A new chapter in the geological story commenced when tropical seas covered the eroded land surface and laid down thick layers of limestones, mud and sand deposits. This period coincided with the rapid evolution of land plants which grew profusely on the flat swampy deltas that stretched over most of Britain, and eventually produced the coal seams previously mined in west Cumbria and other parts of the UK.

The period of coal formation was terminated by further earth movements which re-elevated the Lake District and the Pennine ridge, depressing the Eden Valley in a basin between these highlands. At this time the region had a desert climate. Much of the limestone cover was stripped away by erosion, confining it to a rim around the core of older slates and volcanic rocks. Thick desert sandstones accumulated in the Vale of Eden and in west Cumbria from the debris of this erosion, and at the same time deposits of salts were formed by desert evaporation of saline

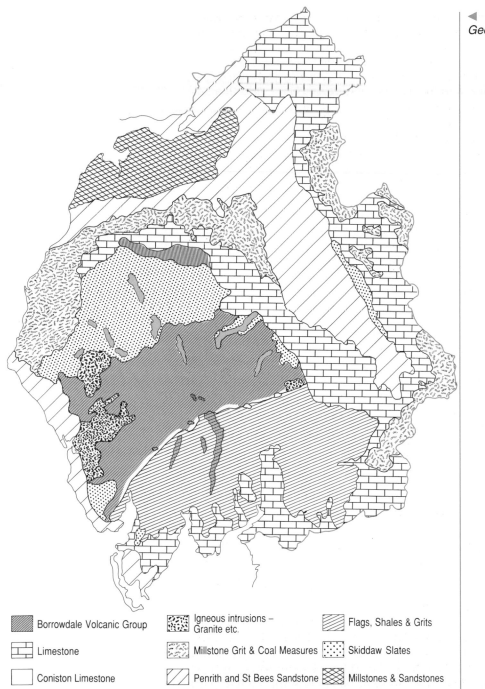

◄ Figure 1.9
Geological map of Cumbria.

▨ Borrowdale Volcanic Group	▨ Igneous intrusions – Granite etc.	▨ Flags, Shales & Grits
▨ Limestone	▨ Millstone Grit & Coal Measures	▨ Skiddaw Slates
▢ Coniston Limestone	▨ Penrith and St Bees Sandstone	▨ Millstones & Sandstones

lakes in low-lying areas, deposits which to this day are being extracted in
the Vale of Eden.

There is little evidence of the geological story over the following 200
million years until another period of earth movements resulted in the final
stages in the moulding of the British Isles, including further uplift of
Cumbria. The area was still warmer than it is today, though humid rather
than arid, and the region would have been covered by temperate
vegetation with gentle streams and rivers cutting V-shaped valleys

between rounded hills. This is what the area would look like now had it not been for the next process of transformation – **glaciation**.

Q   The description of the formation of the rocks of Cumbria shows not only large vertical movements of the land, but also rocks formed in desert, humid tropical and temperate conditions. How could this have occurred?

A   There are two possible explanations. There might have been large changes in world climatic conditions *or* the rocks of Cumbria might have been formed in different positions on the Earth's surface. Surprising as it may seem, current scientific opinion is that all the land masses of the world have moved horizontally as well as vertically. During the formation of its rocks, Cumbria has moved from the mid-latitudes of the southern hemisphere, crossing the Equator and desert zones. This surprising aspect of geological history known as 'continental drift' is discussed further in Chapter 5.

Glaciation

The realisation that the landforms of the Lake District had been created by ice dates back to observations made by William Buckland, Professor of Geology at Oxford University from 1813. He noted features such as the steep-sided and flat-bottomed valleys, and even deep scratches on exposed rock surfaces, which could not be accounted for by geology or erosion by water. As a devout Christian, he was reluctant to accept the possibility of such a radical occurrence as a period of erosion by ice. However, in 1840, he brought the pioneer Swiss glaciologist Louis Agassiz to the Lake District and Agassiz soon demonstrated that the landforms were exactly consistent with the existence of valley glaciers similar to those still present in the Alps. Subsequent study has confirmed that Cumbria has, like much of Britain, been affected by a succession of glacial periods over the last two million years, separated by interglacials with temperatures similar to those of the last ten thousand years.

Recent research on Antarctic ice cores, ocean sediments and deposits in caves has shown that there were many alternations between warmer and colder periods during the period of Ice Age. Indeed, because many of the warmer periods were as warm as and longer than what we call postglacial times (time since the last glaciation), it is probable that we are now in an *inter*glacial period.

At its maximum extent, an ice sheet covered the UK almost as far south as the Thames, as part of a sheet covering northern Europe and extending into the Atlantic west of Scotland and Ireland. At that stage erosion tended to flatten landscapes, and ice was more obvious as the cause of deposition of rock debris, including boulder clay – the mixture of clay and stones of various sizes that covers much of lowland Britain and is apparent in parts of the Pennines. The distinctive features of the Lake District and other areas like Snowdonia were produced in less cold periods, when the highlands of the UK produced valley glaciers but were not covered by continuous ice sheets.

The starting point would have been a relatively smooth dome, probably with a radial pattern of shallow valleys cut into it. As climate worsened, more snow would have fallen and it would have survived longer into the summer, particularly on north-facing slopes. As the glacial period set in, the snow remained year-round and began to accumulate. As the layer grew thicker it would also compact under its own weight into a granular form of ice. In some hollows the pressure grew large enough to start this ice flowing downhill to form a glacier.

Valley glaciers produced many features now familiar to Lakeland walkers (Figure 1.10). At the upper end of many glacial valleys is a natural amphitheatre called a combe or cove in Cumbria (a cwm in Wales, a corrie in Scotland and a cirque in the Alps) with a very steep back wall where the ice pulled away pieces of rock. It often has a ridge on the downstream side, now enclosing a lake called a tarn. Where two adjacent valleys each had combes eating back into the ridge between, the ridge was progressively sharpened to produce features known in English as edges and to Alpinists as arêtes. Where several combes surrounded a peak, these edges would come together in a sharp summit typified in the Alps by the Matterhorn, but represented on a smaller scale by Great Gable. The peak of Helvellyn approached by Striding and Swirral Edges and containing the combe occupied by Red Tarn is a perfect example of this complex of glacial features.

Along the length of glacial valleys, the extreme power of downcutting by ice transformed previously V-shaped river valleys to the typical U-shaped glacial valleys. The difference in power between main ice stream and small tributaries would leave the latter as 'hanging valleys' ending at the shoulder of the U, often with waterfalls or cascades to the floor of the main valley. Many of these features are named 'forces', as in Aira Force, and the resulting gully a gill or ghyll. Unlike rivers, glaciers can flow upwards for short distances, sometimes leaving ridges across the valley which now hold back lakes.

It should be apparent from this description that much of the drama of the Lake District – the lakes themselves, the steep rock walls above tranquil tarns, the sharp edges and pyramid peaks, the waterfalls emerging from hanging valleys and tumbling to the flat valley floors – result from the coexistence of a temperate wet climate and glacial landforms. As such, the landscape is far from being in **equilibrium** and is therefore undergoing rapid change. Given sufficient time, the lakes will silt up, the gills will widen and coalesce to change the U-shaped valleys to Vs and the summits

◀   *Red Tarn between Swirral and Striding Edges.*

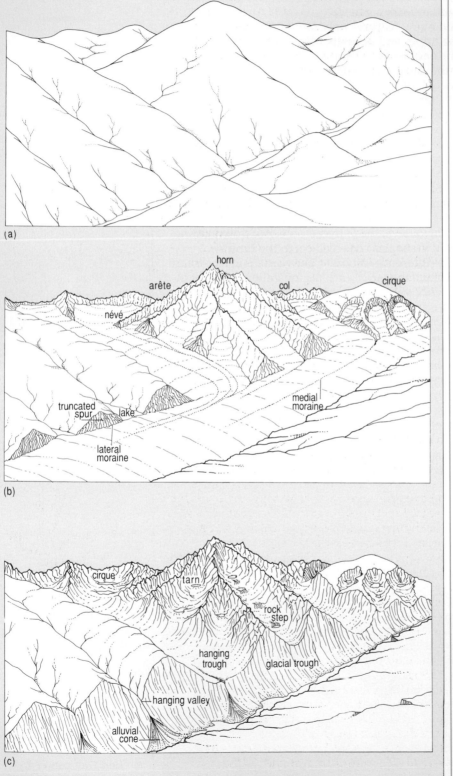

◄ *Figure 1.10*
Glacial features:
(a) before glaciation;
(b) during glaciation;
(c) after glaciation.

will be smoothed and rounded, to return the landscape to that appropriate to humid conditions. Under natural conditions, soil and vegetation would play a strong role in this process, stabilising the gentler slopes, but leaving steep slopes open to frost and gullying and hence to more rapid erosion. The importance of vegetation and soil are demonstrated by the sharply accelerated erosion which occurs where too many sheep or walkers wear away the vegetation and open the ground to the effects of rapid run-off of heavy rain.

Vegetation: natural and unnatural

After the last retreat of the ice, there was a period of colonisation of the area by plants and animals appropriate to the warmer and wetter conditions. However, this period was only about 10 000 years before human activity began to have detectable effects, so even the natural vegetation was rather recently established. Much of the information about this period comes from studies of pollen grains found in lake sediments, supplemented by evidence of remains from peat bogs. In both these situations, hardly any oxygen is available for decomposition, and organic materials survive for millennia – especially pollen, which is both resistant to decay and a good indication of the major kinds of vegetation in the area.

When the ice retreated it left a landscape without soil: there would have been large areas of bare rock, slopes of scree, patches of boulder clay and areas of sand and silt. The spread of vegetation depended on adaptability to climate, mobility of seed and ability to survive on rock or mineral debris. The process would have been one in which early arrivals, usually mosses and lichens, would have produced changes (organic debris and soil development) which allowed more demanding species to survive. The eventual result was to cover most of the area with woodland and to create an almost continuous soil cover.

The majority of the area up to altitudes of about 600 metres would have been wooded, with sessile oak as the dominant species. At higher levels, species like birch and Scots pine would have replaced the oak. In the limestone areas, ash, elm and hazel played a more major role, while in wetter areas willow and alder would have dominated. Differences in soil and tree cover would in turn have influenced the smaller plants to produce different communities. On the higher slopes, grasses such as bent or mat-grass, heather, berry plants like bilberry, and dwarf willow would have covered all but the most exposed rock slopes, where only mosses and lichens could survive, and the bogs, where sphagnum moss dominated and curiosities like sundew and butterwort compensated for lack of soil nutrients by catching and digesting insects.

The amount of tree pollen in lake sediments began to diminish from about 4000 BC as Neolithic (later Stone Age) farmers moved into the area. Because they used cereals and animals for food, they began the process of forest clearance, aided by stone axes. This has eliminated the whole of the natural forest cover, with the possible exception of relict woodlands at Keskadale and Martindale. Though considerable areas of broadleaved woodlands persist, they are certainly much altered by centuries of use and sometimes, as in the case of beechwoods, are entirely composed of species imported to the area and planted. If human beings and sheep were kept out of the Lake District, most of the grassland would change first to scrub and then to woodland. In other words, it would develop towards the vegetation cover which is the natural equilibrium response to the landforms and climate.

Conclusion

In conclusion to this subsection on the natural components of the Cumbrian landscape, I want to emphasise three related points.

1   These natural features are characterised by change: even long periods of equilibrium turn out to have been temporary as they are overtaken by dramatic new developments. This dynamism is confirmed in the geological record and in climatic change, and further change in either could have enormous impacts on human society.

2   The natural landscape does not merely provide a platform on which social processes take place. On the contrary, the different aspects interact: geological differences affect landforms and climate (altitude and aspect make high or north-facing areas colder and wetter than their surroundings), soil formation and vegetation.

3   The vegetation we now see is nothing like the natural vegetation, but has been modified by human activity over millennia. Animal life is very easily influenced by human intervention and can affect plants, rates of erosion and hence the productivity of land and water. Some of these effects are obvious, others are more subtle, but all need to be understood if human society is not to cause unintended, and often damaging, environmental change.

Q   Pressure from the feet of large numbers of walkers has removed grass cover and allowed erosion of paths on Catbells. Think back over the subsections on glaciation and vegetation and say how many kinds of disequilibrium are explicitly or implicitly involved.

A   Three:

1   The erosion of paths results from a disequilibrium between walkers, grass cover, climate and slope. Removal of grass allows running water to remove soil and create gullies.

2   The grass cover itself is a result of humans and sheep clearing the natural woodland. Without continued management, Catbells would in time change to woodland.

3   The ridge on which Catbells occurs is a product of glacial erosion and is out of equilibrium with the present climate. Arguably, by accelerating erosion of the ridge and deposition of debris in Derwent Water, footpath erosion is accelerating change towards equilibrium.

It should be clear from these comments that the idea of equilibrium describes the balance between forces and relates to the rate of change, but is not in itself necessarily good or bad. Such evaluations are made by people in terms of their interests and values.

3.3   Human society and Cumbrian environments

Introduction

It is already clear that human society has had a pervasive effect in modifying the vegetation cover in Cumbria. In turn this has some effects on soil erosion and deposition in lakes and estuaries. A more dramatic impact on the landscape is created by mining and quarrying and an almost total transformation is exerted by urbanisation and industrialisation. These impacts can be traced far into prehistory: the quarrying and manufacture of stone axes in Langdale was the first known industrial use of the area and the source of much greater capacity to clear woodland, both locally and also in the wide area over which Langdale axes were distributed. This was an early step in a process of technological development which has given modern society the ability to transform environments, for better or for worse.

The last section showed a major environmental division in Cumbria between uplands and lowlands. The uplands were not only higher but steeper, wetter, windier and colder – no wonder that Westmorland was for long the most sparsely populated area in England. The physical problems of development were also compounded by the fact that this was politically a frontier from the time of the Romans to 1603. The border between England and Scotland ran through Dunmail Raise and added many skirmishes and raids to the problems of living in the uplands. Small wonder that the early settlement and major market towns were in the lowlands: Carlisle, Cockermouth, Whitehaven and Kendal had half the area's population as late as 1851. Though the upland areas had by then been settled, the populations tended to look out towards their nearest market town and had little conception of being a single district. Indeed, the area was divided between the counties of Cumberland, Westmorland and Lancashire until the creation of Cumbria county in 1974.

Agriculture

The leaders in agricultural improvement in the medieval period were the abbeys, such as Furness, Calder and Cartmel. They began to drain and lime their arable land in the lowlands and to develop their large upland holdings. Trees were cut in large numbers, to supply wood for building and fuel to smelt minerals as well as to clear land for pasture. Sheep and cattle from the higher land were often moved to the lower land for fattening. In many upland areas, the pasturing of sheep prevented regeneration of woodland and accelerated the change from woodland to grassland and heath.

After the dissolution of the monasteries, leadership in agricultural improvement tended to pass to the larger estates, which had the capital to invest in drainage, machinery and improved varieties of stock. Up to the eighteenth century, production was for local markets and therefore included both cereals and animal products. The most productive areas were the lowlands of the Solway Plain and Eden Valley, with Carlisle as the service centre and major market. As road, rail and sea transport grew more efficient in the nineteenth century, Cumbria was increasingly seeking to send its produce to more distant urban markets but was also exposed to competition from less harsh areas in the UK and later abroad. The result was a sharp reduction in sales of arable produce and a growth in sales of animal produce: mutton and beef, dairy products and wool. This shift was encouraged by consolidation and enclosure of agricultural holdings, which affected not only the lowlands and valleys, but also produced the intricate pattern of dry stone walls which reach high up the fells. This allowed an increase in numbers of animals grazed in the uplands, many of which subsequently moved to lowland fields or farms for fattening. The lowlands also imported Scottish, and later Irish, beef cattle for fattening. The dominance of livestock helped Cumbria resist the agricultural depression brought about by cheap cereal imports from America and from the 1870s onward – rather better than did arable areas in the south-east of England.

The overall pattern of Cumbrian agriculture has developed relatively slowly, though intensification has been faster in the lowlands. In recent decades the survival of sheep and cattle farming in the uplands has been sustained by UK and EC subsidy, most recently the ewe and lamb premiums, and hill livestock compensatory allowance. Payment per head of livestock has tended to increase sheep numbers and promote overgrazing, with particular damage to heather moorland, reductions in wildlife and increases in soil erosion. During the late 1980s, reductions in subsidy for cereal growing and dairying have led to a trend to increase sheep numbers in lowland areas of south-east England, a process which could render upland sheep rearing uneconomic. There is now a strong possibility of dramatic change in upland farming, unless current initiatives to redirect subsidies toward environmental management are taken up very energetically by the Ministry of Agriculture, Fisheries and Food (MAFF) and the Common Agricultural Policy.

One of the possible results of a reduction of sheep farming will be increased pressure for afforestation. At present, about 10% of the Lake District is under forest (as is the rest of the UK), of which half is broadleaved (especially oak and birch) and half coniferous (spruce, larch, pine, fir). One of the ironies of conservation is that afforestation arouses intense opposition even in an area which would be naturally forested. The major reason was the planting by the Forestry Commission in the 1920s and 1930s of large blocks of moorland with non-native conifers, most controversially around Ennerdale. These had dramatic impacts on landscape and wildlife. So great was the outcry from walkers and conservationists against 'alien conifers' that in 1936 the Forestry Commission entered an agreement with the Council for the Preservation of Rural England not to plant in the centre of the Lake District. This agreement continues to be vital because the planning authorities have no control over afforestation. The Lake District Special Planning Board has a forest policy designed to encourage appropriate planting of broadleaved trees in the central parts of the Park, with conifers relegated to the lowland fringes, but the economics of timber production tend to mean that hardwoods are planted mainly for amenity purposes and with grant aid, so pressure for commercial conifer plantations remains strong. It remains to be seen whether more sympathetic planting schemes, more mature plantations and provision of camp sites, nature trails and visitor centres will make commercial coniferous woods more acceptable to the general public.

Mining, quarrying and industry

As mentioned earlier, these activities have a very long history, going back to the Neolithic stone axes. Quarrying for local use would have developed in parallel with the process of settlement. However, commercial mining and quarrying really only started in the sixteenth century and had their heyday in the nineteenth, bringing about large-scale impacts on the environment. During this period a dozen different minerals were exploited: for example, graphite (a form of carbon) in Borrowdale (used in the manufacture of the world first 'lead' pencils in Keswick in 1558 – an industry which still survives there), copper at Coniston, silver at Glenridding, lead at Alston, tungsten ore at Carrock Fell, and gypsum and barytes (which are still being extracted) at Newbiggin and Coledale. However, the most dramatic impacts resulted from the extraction of coal along the west Cumberland coast and iron-ore in and around Furness.

The most striking impacts of mining were the direct and indirect results of the discovery of what were then the world's largest known deposits of haematite iron-ore at Park and Hodbarrow in 1850 and 1856. This ore was 60% iron – far richer than other British iron-ore deposits, and not contaminated, as were others, with phosphorus. Up to that time the richness of the haematite ore had supported small-scale smelting using charcoal, but had proved a disadvantage to large-scale local industrial development because the coal mined in Cumberland was too soft to make coke for smelting – it was cheaper to move the ore to coalfields like Staffordshire (by rail) and south Wales (by sea) than to bring coking coal to Cumbria for smelting. However, the size of the new deposits prompted the formation of the Barrow Haematite Steel Company and the construction of the Barrow Ironworks in 1859. The timing was perfect: in 1857 Henry Bessemer, with the aid of Robert Musket, had perfected his converter, which revolutionised the manufacture of steel, and which relied on non-phosphoric iron-ore (Figure 1.11). The coincidence of the ore and the technology made Barrow steel manufacture a dramatic success: by 1889 Barrow was producing one-fifth of England's output of pig iron, and the population of Barrow grew from 3135 in 1861 to 47 259 in 1881. Iron and steel works also proliferated in west Cumberland, but tended to lag behind Barrow in the peak years.

However, the boom years were short lived. By 1878 the Gilchrist Thomas process had solved the problem of steel-making with phosphoric ore, and Cumbria lost ground to more centrally located industrial areas. Profits declined from the 1870s, production was down to a tenth of the national output by 1900, and there were many closures as plant size increased. The legacy in the twentieth century has been the scars of mining and spoil heaps, derelict industrial buildings and the existence of industrial towns and villages whose *raison d'être* has withered as the century

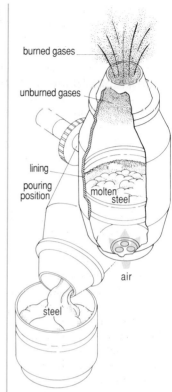

▲   Figure 1.11
The Bessemer converter.

▲   Wellington Pit, near Whitehaven, in about 1915.

progressed. For example, iron mining in Furness was gone by 1931 (though Hodbarrow persisted until 1968) and most of the Cumberland coal mines have closed. Since the decline of mining, quarrying has reasserted itself and there are now large granite and limestone quarries at Shap and Kendal and eight active slate quarries within the National Park, as well as small quarries for building stone. Open-cast coal mining has been added as a new impact on environmental quality.

Since the decline of Victorian heavy industry, Cumbrian industry has come to be divided between the depressed towns of the west coast and the more prosperous east. The west coast still contains large plants, including Vickers' submarine yard at Barrow, major chemical works and Sellafield, but the attempt to establish the motor industry in the 1970s failed to survive in spite of the construction of the A66 road. In the east, a more diversified industrial structure persists, with roots going back to the industries of the early nineteenth century. Then, local raw materials such as wool, corn and wood supported widespread and diversified industries including the manufacture of textiles, bobbins, baskets, paper and gunpowder. The main concentration of industry is now Carlisle, with traditions of textiles, food processing and light industry and a long history as market town and service centre, including provision of financial services to a wide area of northern England and south-west Scotland. On a lesser scale, Kendal and Penrith share this diversified industrial base and the advantage of good road and rail links to north and south. The legacy of nineteenth-century boom plus twentieth-century depression has left towns like Barrow, Whitehaven and Workington with urban problems reminiscent of the inner cities of the large industrial conurbations, while Carlisle and Kendal share in the more buoyant economy and growing service industries of market towns in other parts of the UK. Towns in the Lake District itself, notably Windermere, Ambleside and Keswick, are largely dependent on tourism.

Tourism

It is one of the paradoxes of history that the opening up of Cumbria for large-scale mining and industry was almost simultaneous with its opening up for amenity uses – tourism and homes for outsiders. Most guide books date amenity uses back to the first edition in 1810 of *Guide to the Lakes* written by William Wordsworth (1770–1850). This was indeed a step up in the level of publicity, but was building on a process developing since about 1750. The book was partly derivative of Father West's guide of 1778, which was a remarkable piece of 'social technology', prescribing a detailed itinerary and even laying down exact 'viewing stations' from which the 'Lakers' would see absolutely the best views of lakes and hills. Well-equipped tourists would even use a 'Claude glass' (a concave mirror with a frame), which required them to stand with their backs to the prescribed scenery and look at its reduced reflection. While this may seem a rather timid and artificial way of communing with nature, one must remember that only a few decades before even a seasoned traveller like Daniel Defoe (1661–1731) had found the area to be ugly and forbidding and best avoided. The imposition of formal 'prospects' on the lakes was a parallel to the aesthetics of the eighteenth-century landscape park, with formal avenues and picturesque ruins. The idea of wild places as spiritual solace or even physical recreation was only just evolving as part of a Romantic reaction against the ugliness and squalor of newly industrialising Britain. Poets like Wordsworth, Robert Southey (1774–1843) and Samuel Taylor Coleridge

(1772–1834) used the Lake District as an inspiration for this new Romantic view of Nature (Figure 1.12), and also pioneered the notion of hill walking as a recreation. However, they found themselves opposing the mass tourism and residential development which their writings did so much to kindle.

Even in the first edition of his *Guide*, Wordsworth devoted a good deal of space to denouncing inappropriate buildings and planting. These were some of the first signs of a trend which was to bring successful manufacturers to settle in the Lakes in retirement or even in second or holiday homes. The resulting Victorian villas and mansions seemed to Wordsworth utterly out of keeping, though today's visitors to Windermere may well prefer them to more recent building. His protests stepped up when railway building started. At first this had concentrated on coastal routes linking the growing industrial towns of the Cumberland coast to Newcastle and later to London. In 1847, however, a railway was built to Windermere primarily for tourists, bringing the lakes within eight hours of Euston station in London. This was the harbinger of fifty years of growing impacts of industry, settlement and tourism on the Lake District, which stimulated new ways of thinking about the proper relationship between society and environment.

Development and conservation

Wordsworth laid the foundations of a conservationist approach in his *Guide* and also in his letters to the *Morning Chronicle* in 1844 protesting against the proposed Windermere and Kendal Railway. He identified the Lake District as 'a sort of national property' in which profit-making should be resisted in favour of 'moral sentiments and intellectual pleasures of a high order' – which, however, seemed to apply only to a select few. His arguments seemed to gain little ground in the face of the national dedication to *laissez-faire* principles and the pursuit of economic growth. However, thirty years later, renewed opposition was stimulated by new proposals to extend railway lines and Manchester Corporation's bid in a private parliamentary bill to use the lake of Thirlmere as a reservoir to

One impulse from a vernal wood

May teach you more of man,

Of moral evil and of good,

Than all the sages can.

Sweet is the love which Nature brings;

Our meddling intellect

Misshapes the beauteous forms of things –

We murder to dissect.

▲ Figure 1.12 *Extract from William Wordsworth's* Lyrical Ballads, *'The Tables Turned', lines 21–8, first published in 1798.*

supply pure water to the industrial cities of Lancashire. A conservationist movement began to grow, inspired by Wordsworth and the wider critique of industrial society by John Ruskin (1819–1900), and was formalised as the Thirlmere Defence Association, which had many aristocratic and intellectual supporters. Both sides began to appeal to a wider need for recreation – conservationists stressing continued access to the lakes in their natural state, proponents of development arguing that Thirlmere would become a 'people's park' and that the building of Gothic villas by 'aesthetic gentlemen' would be restrained. A crucial precedent was set: previously, private parliamentary bills could only be opposed by people whose property would be affected directly, but for Thirlmere a wider public interest was admitted. The bill was defeated in 1878, but reintroduced in 1879 and passed, so Thirlmere became a reservoir.

This failure for the conservation lobby, coupled with a proposal to build a new railway from Keswick to Buttermere, prompted the formation of the Lake District Defence Society, which harnessed the energies of Canon Rawnsley of Crosthwaite, as well as the Thirlmere activists. Early successes in stopping railway developments, plus energetic promotion, soon built up a membership of 600. Only 10% were resident in Cumbria: half the membership was from London and industrial Lancashire, including many professors (some from the USA), clergy and public school teachers. The Society suffered no defeats between the mid-1880s and 1914, by which time the pressures for development had become less extreme. Increasingly, local activists, including landowners, could work through contacts in county and central government to ameliorate proposals for roads, telegraph or electrical supplies. This national network, which included the Lake District Defence Society, Octavia Hill (who had created charities to provide small urban parks as well as working class housing) and the Commons Preservation Society, was crucial in facilitating the formation of the National Trust as an organisation to purchase key areas and monuments, and some of its earliest acquisitions were around Derwent Water.

To summarise this section is a quotation from a study mentioned earlier:

> The issues posed by the pressures to open out and develop central Lakeland played a significant part in the retreat from *laissez-faire* economic doctrines in the later nineteenth century, and it is here, perhaps, that Cumbria made its most significant impact on the history of the nation at large.   (Marshall & Walton, 1981, p. 219)

State intervention and the National Park

The retreat from *laissez-faire*, however, was a long and slow process, and was far from complete even at its maximum. It produced new problems as well as new ways of reducing old problems. On the one hand, growing local and central government power provided substantial checks on development proposals by private companies (a process paralleled by the comparative weakness of development interests in the depressed 1920s and 1930s). On the other hand, growing state intervention produced powerful organisations like the Forestry Commission, the Central Electricity Generating Board, the Department of Transport and the UK Atomic Energy Authority, which posed even more irresistible development pressures. The concentrated power of these organisations, allied to the inexorable growth in car ownership, has continued to produce substantial development pressures in Cumbria, as evidenced by the current issues discussed in Section 2 above. The constant retreat in the face of development has led conservationists to look again at the concept of the National Park as it evolved in Cumbria and the UK.

As already mentioned, the idea of the Lake District as 'a sort of national property' dates back to Wordsworth. It was developed into a wish for a national park by the 1880s, influenced by the establishment of the first National Parks in the USA as well as by the experience of the Lake District Defence Society. However, the Lake District was not an undeveloped area owned by the government, as was Yellowstone or Yosemite, but an area with a substantial population, a long history of occupation and almost entirely in private ownership. Moreover, the ideas of nature conservation and outdoor recreation were far less central in British culture than those of wilderness and the frontier in the USA. No wonder it took so long to turn the idea of a national park system into a reality, nor that that reality was a typically British compromise, applying as it did only to England and Wales but not to Scotland or Northern Ireland. The detail of the official reports and pressure group activities which led to the National Parks and Access to the Countryside Act of 1949 and the establishment of the Lake District National Park need not concern us here. However, the goals are important, as is the discrepancy between the goals adopted and the limited powers available to pursue them.

The very name of the Act signals one of the clashes in principle: the parks were intended not only to conserve valued landscapes, but also to promote public access. The fact that these principles might conflict was only realised later. It was accepted from the start that because the landscape to be conserved was a socially modified landscape, it was essential for the park authority to work with existing owners and users, such as farmers, foresters and hoteliers, rather than excluding them. This tendency was compounded by the simultaneous establishment of the Nature Conservancy (now English Nature) to define and acquire the country's most significant natural areas as a system of National Nature Reserves, from which other uses, including public access, were to be excluded. These nature reserves were far smaller than the National Parks (111 of them occupy only 0.25% of the area of the UK, whereas the 11 National Parks occupy 10%), and were maintained essentially for their scientific interest. Against the Nature Conservancy's (admittedly small) budget for purchase of reserves, the national park authorities were essentially required to operate using the same planning powers as those available to all local authorities after 1947, i.e. the ability to control development of new buildings and changes of use of existing ones – except in the cases of agriculture and forestry! Until 1974, the Lake District Special Planning Board was one of only two national park authorities which overlapped more than one county, and hence was at least an autonomous organisation rather than a department within a larger authority. It has undoubtedly played an important role in guiding the development of the Lake District in the years since 1951, but has often had to rely on persuasion rather than legal or economic powers. Even its central power of development control is subject to appeal to the Secretary of State for the Environment. At the height of the building boom of 1987–88, planning applications were running at 100 per month, and though 80% were refused, 125 were the subject of appeal (up 61% on the previous year) of which 59 were allowed (as against only 30 the previous year). Small wonder that UNESCO concluded that the available controls were not strong enough to safeguard the Lake District as a natural World Heritage Site.

Summary

At the end of the 1980s, the Lake District remains Britain's premier National Park. But the park itself is subject to a variety of pressures for change, of which the major current issue is tourist pressure, bringing physical erosion to footpaths, traffic congestion (and proposals for road improvements), and expansion of settlements. A new threat to the landscape comes from changing farm subsidy levels, with the attendant possibility of the decline of pastoral agriculture and reversion to scrub or planting for commercial forests. Moreover, policy for the National Park must be related to planning policy for the county as a whole, where economic decline has left a legacy of old industrial towns unable to offer the amount or quality of jobs needed by their inhabitants. In this situation, the rival claims of environmental conservation and economic development are highlighted, not least by the Faustian bargain of 15 000 well-paid jobs at Sellafield in exchange for the risks of nuclear pollution. For our purposes, Sellafield has the advantage of dramatising the political and economic problem of growth versus conservation and also of prompting extensive environmental monitoring and consequent insights into natural processes. I will turn to these processes in the last section of this chapter.

Q   What aspects of the physical environment of Cumbria have been used
    as resources by society?

A   At least three different kinds of resource have been emphasised:
    (1) deposits of minerals, e.g. coal, iron, gypsum laid down by past
    geological processes; (2) the soil, climate and vegetation which have
    provided the inputs to agriculture; and (3) the appearance of the
    landscape which has been the basis for poetry and tourism. Resources
    are very much the product of the relationship between environment
    and society.

Q   What have been the main impacts of society on the physical
    environment in Cumbria?

A   The most pervasive impact has been the transformation of the
    vegetation by agriculture, the most intense the scarring by mining,
    quarrying and heavy industry. Urbanisation and transport routes have
    dramatically changed many areas. Other impacts may be less obvious
    but important nevertheless: air pollution in the past affected plant
    growth in the Furness district, and water pollution by industry, sewage
    and farm chemicals could be demonstrated. Radioactive pollution is
    widespread, as will be shown in Section 4.

Q   How have competing demands on the Cumbrian environment been
    managed?

A   The interaction between competing demands has passed through
    different stages: in the nineteenth century market forces decided
    which activities would occur, though the Thirlmere case showed the
    activity of municipal government and the need for Parliamentary
    approval of major projects. In the twentieth century, various state
    bodies began to play roles, from provision of particular services to
    regulation through the planning process. The continued, indeed
    enlarged, amount of pressure-group activity suggests that state
    intervention has by no means solved environmental problems. The

complex processes that determine environmental outcomes will be an ongoing topic in later chapters.

Q   Give examples of the use of technology to create or solve environmental problems.

A   The use of technology has been ubiquitous, from stone axes to nuclear reprocessing, but the most dramatic impacts were probably exerted by steel technology in the past and transport technology in recent years.

Activity 6

Think about the county, city or region you live in. In what ways do the local industries draw on local resources and what impacts do they have on environmental quality? Does your monitoring of the local media suggest that there is any tension between pressures for development and the conservation of natural or built environment? Does regulation through the planning system and/or political activity achieve a satisfactory balance between exploitation and conservation? Are there any active pressure groups arguing for particular points of view?

4   Radiation monitoring and environmental processes

4.1   Why study Sellafield?

The release of radioactive substances into the Cumbrian environment has two aspects which make it a suitable topic for an introduction to an environment course. First, scientists' ability to detect even very small amounts of these substances makes it possible to trace how they move through a variety of environmental processes, in water, air, soil, vegetation, animals and in human bodies, highlighting the existence of complex interrelationships. Second, the escape of radioactive materials from Sellafield, both accidental and deliberate, has stimulated an intense debate about the effects on human health. The fact that such a vital debate is as yet inconclusive emphasises that human society often tends to engage in potentially damaging activity without being able to assess the risks. Indeed, the extent of environmental monitoring in Cumbria is in substantial part a result of anxiety about radioactive pollution, as half a dozen official bodies and several pressure groups seek evidence for what is really happening. This evidence, and the arguments based on it, are the raw material of this section.

The question to be addressed is simple: Do the benefits of Sellafield justify the risks? The answer is more difficult: the claimed benefits have changed over time, the benefits have all been challenged, and the risks are difficult to assess. Typically, supporters and opponents of the plant use different evidence, lines of argument and criteria of success or failure. In presenting the different arguments, I try to take a dispassionate look at their strengths and weaknesses and in particular to highlight points where better information is needed to substantiate or refute the argument. This is because the purpose of this section is not just to analyse the benefits and risks of Sellafield, important as they are, but to use this case to show the need both for better understanding of the environmental impacts of pollutants and for better ways of deciding whether potentially dangerous activities should be started or allowed to continue.

The distinctive function of Sellafield has been reprocessing spent reactor fuel. The goal of reprocessing – and hence the claimed benefits of the plant – has changed over time. Initially it was to extract plutonium for use in bombs; increasingly, the emphasis shifted towards recycling uranium plus stabilising and storing nuclear waste. More recently, reprocessing of waste from abroad has been seen as a lucrative export industry. As all of these have been challenged, the final redeeming benefit of the plant has been recognised as the provision of work in a depressed area.

The risks of the Sellafield plant range from routine exposure of workers and discharges to the environment through to possibly catastrophic accidents. The range is from a high probability of small doses of radioactivity received by certain people to a very low probability of high doses in a serious accident. At both ends of this continuum, there are sharp differences of view between official appraisals and public responses. This is most extreme for catastrophic accidents where the official appraisals suggest probabilities very close to zero; they therefore conclude that such accidents will not occur, in spite of accidents in the past and at other plants. Members of the public, in Ireland as well as the UK, are more likely to focus on the size of the potential catastrophe and to be very concerned about the risk even if the probability is extremely low. In recent years it has been disclosed that there have been substantial numbers of accidents in the plant, in most of which there was no release of radioactivity to the surrounding area, but none of them compares in seriousness to the worst recorded accident – the Windscale fire of 1957.

In view of the difficulty of calculating and interpreting the probability of serious accidents, the discussion below will focus on the risks from known discharges.

Even this involves difficulties as different interest groups have taken different routes, which have led to different conclusions. One line of argument is to monitor the radioactive discharges from the plant and/or the amounts of radioactive material in water, sediment, soil and vegetation around the plant, and then compare these levels with levels of natural background radiation and with official estimates of risk derived from more than half a century of scientific and medical research into the effects of radioactivity. The second line of argument is to measure directly the incidence of illness known to be induced by radiation, notably leukaemia, and to compare areas near Sellafield with other parts of the UK. Ideally, if all the steps of the argument were correct, these two lines of argument should corroborate each other. In practice, at the time of writing, they seem to lead towards different conclusions – the first showing Sellafield to be safe and the latter suggesting that it may not be. This implies that some of the steps in one or both arguments may be misleading. Perhaps unsurprisingly the operators of Sellafield have always adopted the first line of argument, whereas opponents, who include the local group Cumbrians Opposed to a Radioactive Environment (CORE) and international groups like Greenpeace and Friends of the Earth, have long argued that discharges were too high.

Because the two lines of argument select from overlapping sets of information, the most convenient way of structuring this section is as follows:

- the natural level of radioactivity in the environment;
- Sellafield and its discharges;
- other sources of radioactivity in Cumbria;
- incidence of leukaemia in west Cumbria and elsewhere;
- summary.

4.2   Radioactivity in the environment

Radiation occurs in a variety of forms, the majority of which are naturally occurring and have positive value as well as posing certain **hazards** (Figure 1.13). The flow of heat and light from the sun is essential to life on earth – but exposure to sunlight also involves exposure to ultra-violet (UV) radiation which can cause painful sunburn or even skin cancer. Radio waves, microwaves and x-rays are all forms of radiation used beneficially in modern society, but in the last two cases care has to be taken to shield people from unwanted exposure. The forms of radiation which cause apprehension about nuclear waste include gamma rays, which are similar to x-rays, and also alpha and beta particles which are emitted from unstable atoms as they break down into simpler forms, decaying with a characteristic half-life (time for half the amount to decay) for each radioactive substance.

Separate measures are used for the *activity* of radioactive substances (the amount emitted), the amount *absorbed* and the biological *effect* of the amount and type absorbed. Details of these measures are given in Box 1.1.

The pie diagram in Figure 1.13 shows the estimate of the sources of the total radiation exposure of the UK population, as calculated by the National Radiological Protection Board (NRPB). The average exposure is about 2.6 millisieverts (mSv) per year, of which the large majority is from natural sources – cosmic rays (including x-rays and gamma rays), gamma rays from radioactive particles in rocks and soils, radioactive gases (radon and thoron) seeping into buildings, and radiation from particles absorbed in our bodies from eating, drinking and breathing.

While this may be the average picture, there is a great deal of variability

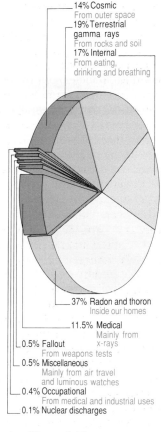

14% Cosmic
From outer space
19% Terrestrial
gamma rays
From rocks and soil
17% Internal
From eating,
drinking and breathing

37% Radon and thoron
Inside our homes
11.5% Medical
Mainly from
x-rays
0.5% Fallout
From weapons tests
0.5% Miscellaneous
Mainly from air travel
and luminous watches
0.4% Occupational
From medical and industrial uses
0.1% Nuclear discharges

▲   Figure 1.13
Source of radiation exposure of the UK population.

Box 1.1   Radiation measurement

The traditional measurement of activity was the curie, but this has been replaced by the becquerel (Bq). Because the becquerel is a very small unit, equivalent to one radioactive breakdown per second, it is often used with the prefix tera, where 1 TBq is a million million Bq.

The amount of radiation absorbed by a body is proportional to its size and inversely proportional to the square of the distance between source and body.

The unit of absorbed **dose** is the gray (Gy), formerly the rad. Figures are usually given as micrograys (μGy), i.e. millionths of a gray.

The biological effect of different kinds of radiation varies by the so-called quality factor (QF), which is 1 for beta and gamma rays but 20 for alpha particles, reflecting the greater biological damage caused by alpha particles.

Dose equivalent=absorbed dose×QF.

The unit of dose equivalent is the sievert (Sv), so for beta and gamma rays 1 gray produces 1 Sv, but for alpha particles 1 Gy produces 20 Sv. (The old unit of dose equivalent was the rem.)

In studies of low dosage radiation, the sievert is inconveniently large, so figures are usually given as millisieverts (mSv), i.e. thousandths of a sievert.

in exposure to natural as well as to artificial radioactivity. The NRPB map (Figure 1.14) shows the variations in **absorption** measured one metre above ground. This would be influenced by rock type, because some rocks, notably granite, contain more radioactive materials than others, and by the amount of rainfall, because this washes radioactive particles (including natural particles) out of the atmosphere. Much greater local variation in exposure can occur, especially to occupants of houses which collect radon (a radioactive gas resulting from breakdown of uranium in rocks), where doses may be twenty or more times the average – well above the doses permitted for workers in the nuclear industry. There is growing concern about the health hazards posed by radon, but as yet no demonstration that high exposure in houses has led to high incidence of cancer or leukaemia.

A further, and probably crucial, source of variability stems from the *form* of exposure to radiation, internal or external. External exposure is likely to involve beta and gamma rays, which travel at very high speeds. Gamma rays are particularly penetrative; for example, they easily penetrate sheet metal and can only be stopped by lead or massive concrete structures. Paradoxically, alpha particles, which are much larger and slower, and can be stopped by a sheet of paper, are biologically the most damaging because alpha emitters can be swallowed or breathed in and deposited inside body organs, where all their energy is concentrated on the few adjacent cells. An analogy would be firing cannon balls and rifle bullets at a wooden palisade: the latter pass through with only localised damage, the former, though less penetrating, smash the timbers to pieces.

Alpha emitters occur naturally. For example, radon breaks down over time to produce polonium, and when this happens in the atmosphere, particles of polonium are washed back to the surface where they may be deposited on or absorbed by plants, then eaten by animals which are in turn eaten by humans. Levels of polonium are especially high in Eskimos and Lapps, who eat a lot of meat from animals that feed on slow-growing lichens, but even in Britain natural alpha emitters including polonium contribute about 10% of average radiation doses.

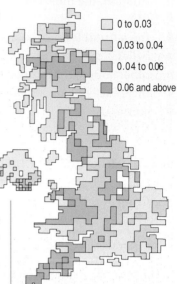

☐	0 to 0.03
☐	0.03 to 0.04
▩	0.04 to 0.06
■	0.06 and above

▲ Figure 1.14
Radiation levels in the UK: the key refers to the absorbed dose rate in air (μGy per hour) measured 1 m above ground

4.3   Sellafield and its emissions

The Sellafield works, consisting of the Windscale reprocessing plant and the Calder Hall nuclear power station, was run from 1948 by the Ministry of Supply, then by the UK Atomic Energy Authority (UKAEA), and from 1971 by British Nuclear Fuels (BNF), primarily for the purpose of reprocessing spent uranium fuel from nuclear power stations. Reprocessing was originally carried out on spent fuel from the 'Windscale piles', and later Calder Hall, to produce plutonium for the secret nuclear bomb programme. More recently it was justified on the grounds that it would otherwise be difficult to store spent fuel rods from Magnox power stations and that there was a need to build up a large stock of plutonium for use in a new type of reactor, the fast breeder (FBR). These justifications were vigorously challenged at the Windscale Enquiry of 1977, but permission was given to build a new reprocessing facility (the Thermal Oxide Reprocessing Plant, THORP) to start reprocessing spent uranium oxide fuel from the new generation of British advanced gas-cooled reactors (AGRs) and imported fuel from pressurised water reactors (PWRs). Since that time, however, the government has ceased to fund the FBR programme because world supplies of uranium are now so abundant that uranium-based reactors appear more cost effective than the FBR. The doubtful need for

▲   *Aerial view of Sellafield.*

reprocessing as part of nuclear electricity generation is shown by the fact that the USA has never used commercial reprocessing as part of its civil nuclear programme. The debate over the value of THORP was revised as the plant neared completion in 1994. In spite of vigorous challenges to its economic viability as well as its safety (Aubrey, 1993), permission was given to start THORP in early 1995.

The issue of reprocessing is an important one because, apart from disastrous accidents, it is easily the most polluting aspect of the UK's nuclear industry and responsible for the great majority of emissions to the environment. The largest amounts have been in liquid form to the sea, but gases and particles have also been released to the atmosphere, most disastrously in the Windscale reactor fire of 1957. This fire broke out in one of the two reactors built solely as sources of military plutonium, and released some 15 000 terabequerels (TBq) of radioactive material (notably iodine-131, which emits beta and gamma rays and has a half-life of only eight days) into the air. This was carried south-east right across the UK and into Europe. At the time the risks were said to be slight, though milk was dumped for some weeks. A later official report for the NRPB (Crick & Linsley, 1983) estimated that the accident would eventually cause 260 thyroid cancers, of which perhaps 10% would be fatal. However, this is only one or two per cent of the death rate in the UK from thyroid cancer, so there is no way of telling whether a particular death resulted from the Windscale fire or not.

Although there have been other accidents involving contamination of air and/or soil, the largest amounts of radioactive materials have been discharged to the sea. These are subject to maximum limits and are closely monitored at source and while moving via various environmental pathways. The figures published by BNF (Figure 1.15) show that the amounts of beta emitters are much higher than those of alpha emitters, and that both peaked in the early 1970s and have been very small since the mid-1980s.

Critics (notably Taylor, 1986) argue that the technology was available from the 1960s to reduce emissions to something like current levels, and indeed that it was used at reprocessing plants in other countries at relatively little extra cost, so that sea dumping from Sellafield became in effect, and perhaps in intent, an experiment which made the Irish Sea the world's most radioactively polluted sea. Proponents of the nuclear industry argue that doses to members of the public are below the limits set by the International Commission on Radiological Protection (ICRP) and hence safe. It certainly is the case that discharges are regulated by the Ministry of Agriculture, Fisheries and Food (MAFF) and closely monitored by BNF, MAFF and other bodies.

The Sellafield pipeline is monitored continuously to produce aggregate figures for 'total alpha' and 'total beta' emitted. Also, periodic samples of the effluents are taken and subjected to laboratory analysis to identify particular components. In 1986 the authorisation was amended to increase the number of specific materials subject to particular limits, perhaps in belated recognition that one beta emitter, plutonium-241, decays with a half-life of only 12 years into the alpha emitter americium-241. The pipeline actually carries more than 20 radioactive substances which have varied physical and chemical properties. The differences are illustrated by two of the principal constituents:

1   Caesium-137 (which emits beta and gamma rays and has a half-life of 30 years) is discharged in a form which is soluble in water and hence behaves relatively predictably, being carried away on the tides and currents and progressively diluted.

2   Plutonium is discharged in a number of forms with different properties, but most are alpha emitters, some with very long half-lives. Most are insoluble in water and were expected to sink to the bed of the Irish Sea and bind to silt particles, which would prevent them being taken up by living things. However, it has been shown that these sediments are moved into estuaries like Ravenglass and Morecambe Bay, so that shellfish which filter food particles from shallow water happen to be one of the ways that plutonium comes to be ingested by people.

The monitoring by MAFF goes beyond measurement of radioactivity in water and silt to establish how people may be exposed to it. Two 'critical groups' are identified, with exposure coming by different 'pathways'. People who live in houseboats, or who spend long periods on mudbanks, are exposed to external doses from radioactive substances bound to clay particles in the mud. Seafood eaters can receive higher doses if they eat a lot of locally caught fish, crabs or shellfish. In a few years in the mid-1980s, the seafood eaters were calculated to receive doses in excess of the International Commission for Radiological Protection's recommended limit of 1 mSv per annum. However, MAFF argue that those calculations assumed a rate of absorption of plutonium in the human digestive tract that is higher than they believe to be accurate. They also point out that doses to seafood eaters have reduced more recently. Critics point out that one reason for reduced

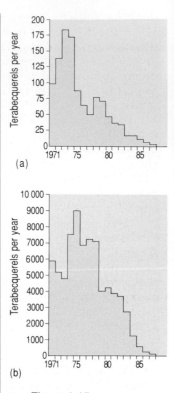

▲   *Figure 1.15*
Radioactive liquid discharge from Sellafield pipeline 1971–87: (a) discharges of total alpha activity to the sea; (b) discharges of total beta activity to the sea.

doses has been reduced consumption of shellfish as a result of recent public alarm about the risks. In 1983 an incident had led to discharges which produced unusually high concentrations of radioactivity on the beaches near Sellafield. Because monitoring showed some objects which could exceed the annual dose limit after a few minutes skin contact, the Department of the Environment advised members of the public to avoid unnecessary visits to the beach.

To the surprise of the scientists, it has also proved to be the case that some plutonium is physically moved inland, probably by the wind picking up foam and water droplets and blowing it inland. The scale of this process should not be exaggerated: Peirson *et al.* (1982, p. 28) estimate that only 0.089 TBq of plutonium from Sellafield has come inland as against 700 TBq of plutonium discharged between 1952 and 1982. It does, however, create another pathway by which alpha emitters may be breathed in or ingested.

Detailed studies of the coastal region have been made by scientists from the Institute of Terrestrial Ecology at Grange-over-Sands. They showed (e.g. Allen *et al.*, 1983) that levels of caesium and plutonium above the tide line were only a few per cent of those below, although very long-lived radioactive substances, notably americium-241, were unevenly distributed across the salt marsh and obviously trapped by certain plants. Radioactive concentrations were lower on saltmarsh grazed by sheep than on ungrazed saltmarsh. Lambs were found to contain more radioactivity than ewes, with concentration in certain organs, especially kidneys. However, lambs were, like the wild birds of the area, found to be far below the safety limits. Overall the 60 hectares of saltmarsh in the Ravenglass estuary contained about 1.5 TBq in sediment with less than 1% of that in standing crops, caesium-137 being the main source.

4.4   Other radioactive contamination in Cumbria: Chernobyl

The potential damage from a major nuclear accident and the unpredictable spread of radioactive materials by climatic events are highlighted by the accident at the Chernobyl nuclear power station in the USSR in April 1986. As well as the need for evacuation of 135 000 people from areas within 30 km of the reactor (Medvedev, 1990), serious effects have been noted in other parts of Europe, including Cumbria. This was the world's most serious nuclear accident, releasing 2 million TBq to the atmosphere, mainly in the form of iodine-131, but also caesium-134 and -137. There were 203 civilians, especially fire crews, who suffered doses high enough to bring on acute radiation sickness. Of these, 45 received doses above 4 sieverts (4000 mSv), the dose which will kill 50% of people exposed. Subsequently, 31 died within 60 days and 24 were totally disabled. No data are available on casualties among military personnel, many of whom were involved in containing the fire. It is now estimated (Sumner *et al.*, 1991) that radioactivity from Chernobyl will cause over 16 000 fatal cancers on the territory of the former USSR over the next 40 years.

Unfortunately for highland areas in the UK, a week after the fire began the cloud carrying radioactive particles passed across the UK and coincided with a period of heavy rain which washed out some of the particles. It is estimated that the average adult in north Wales, Cumbria and south-west Scotland received a dose of 0.27 mSv (Sumner, 1987), about 10% of the annual dose from all sources. However, the main lasting effect has been on sheep. It was expected that the effects of fallout would be shortlived because iodine-131 had a half-life of only eight days, and previous studies of radioactive caesium had shown that it was quickly bound to clay particles in soil and immobilised. The initial restriction on slaughter and movement of sheep which exceeded the safety limit of 1000 Bq per kg (the level calculated as giving keen eaters of lamb an annual dose of 1 mSv) was imposed for only three weeks. This level of radioactivity in sheep has been much more enduring than was expected. The main reason is that upland soils contain little clay but a high proportion of organic material: they therefore fail to bind caesium as do lowland clays. Moreover, whereas the caesium was initially ingested by sheep as particles deposited on vegetation, it is increasingly being taken up by plants from the soil and incorporated into the plants themselves. These organic forms of caesium are more easily absorbed in the sheep's gut and hence incorporated in the meat. In 1995, 70 000 sheep in Cumbria were still under restriction. Fortunately for Cumbrian farmers, the traditional practice of selling upland lambs for fattening on the lowlands provides a period of grazing on uncontaminated pastures; this allows them to eliminate radiocaesium from their bodies and pass as fit for consumption. It remains uncertain how long it will take for the level of radiocaesium in remaining upland lambs to fall to negligible levels, but experience so far suggests that it will be a slow process. Wynne (1992) has argued that discrepancies between official predictions and actual experience has undermined farmers' confidence in science and encouraged speculation that pre-Chernobyl levels of radioactivity in Cumbria were higher than officially recognised.

Activity 7

Box 1.1 on radiation measurement identified different units of measurement for activity, exposure and absorbed dose equivalent. Review these units and find examples of the use of each in the text and illustrations. Using this information answer the following questions.

Q   Was the activity released by the 1957 Windscale fire greater or less than the activity released into the sea in the peak year?

A   Greater. The fire released 15 000 TBq whereas releases to the sea in 1975 included just over 9000 TBq of beta emitters and 80 TBq of alpha emitters.

Q   Which has released the most radioactivity: British Nuclear Fuels Sellafield or Chernobyl?

A   The release from Chernobyl, at 2 million TBq, was vastly greater than that from Sellafield, where the annual releases in Figure 1.15 sum to less than 80 000 TBq.

Q   Which led to the highest doses to Cumbrians: fallout from Chernobyl or eating local seafood?

A   The 'critical group' of seafood eaters received doses over 1 mSv in some years whereas average doses from Chernobyl were estimated at 0.27 mSv.

4.5   Argument and counter-argument

For the first half of its existence, the Sellafield operation was shrouded in secrecy because of its role in military technology. More recently, the growing proportion of work on the civil nuclear programme plus the activities of pressure groups in publicising leaks have increased the provision of information about discharges. As late as the 1977 public enquiry, the official case that Sellafield was safe and that reprocessing was economically vital prevailed in spite of vigorous counter-arguments from pressure groups. A new debate was stimulated by the 1983 Yorkshire Television programme *Windscale – the Nuclear Laundry* and its revelation that the nearby village of Seascale had a rate of childhood leukaemia ten times as high as the national average. The details of the arguments have become extremely complex and are not necessary to the purpose of this chapter. The basic structure of the contrary arguments is enough to demonstrate the points needed in this introductory chapter.

●   the complexity of the responses of environmental processes to human impacts;

●   the complexity of economic, governmental and pressure-group influences on decisions;

●   the existence of an explicit or implicit process of **risk assessment**.

Taken on a national scale, the evidence of leukaemia clustering is itself quite difficult to understand because leukaemia is quite a rare disease, so a cluster usually consists of only a few cases. The Seascale cluster is of five cases where the average national incidence would produce an expectation of only half a case. To complicate the picture, there are other clusters (including Tyneside, Sedburgh and Whittingham in the north of England) which are far from nuclear facilities. Also, although two clusters are near the Dounreay fast breeder reactor in Scotland and the Aldermaston and Burghfield weapons plants in Berkshire, all monitoring data suggest that discharges from these plants are a very small fraction of those from Sellafield, so the problems of explanation are even more severe than at Seascale.

The official case has long been that non-natural radiation doses to the public have always been below the safety levels set by the International Committee for Radiological Protection – a maximum of 5 mSv per annum and a recommended limit of 1 mSv. When elaborating this position in presenting evidence to the Black Committee (which was set up to investigate the leukaemia cluster), the NRPB used BNF and MAFF data on emissions to calculate that the additional radiation dose to Seascale children between 1950 and 1970 would only have been a quarter of the dose from natural sources. They argued that even when using a much higher risk factor than that recommended by ICRP, only 0.7 deaths from leukaemia could be explained by this radiation dose and only 0.1 of these could be attributed to Sellafield. Because of this, they argued that there must be some other cause of the leukaemia cluster in Seascale. The Black Committee recommended further research to reassure the public that everything possible was being done. The result was the formation of the Committee on Medical Aspects of Radiation in the Environment (COMARE) and the commissioning of further research.

The main development in the new research programme was the finding by Gardner (1990, 1993) that childhood leukaemia was statistically related to levels of radiation exposure of men who worked within the plant before becoming fathers. This line of argument was challenged by Evans (1990), who pointed out that farmers, steelworkers and chemical workers suffered similar rates of leukaemia to those experienced by nuclear workers. In spite of this, the parents of two of the Seascale children sued BNF for damages in 1993. After exhaustive evaluation of the arguments put forward by over 50 expert witnesses, the High Court concluded that there was no proof that radiation from Sellafield was the cause of the illnesses of these children. No other definitive explanation emerged; some expert witnesses indicated virus infections as a possible trigger, some witnesses implicated chemicals, and others attributed the cluster to chance (Milne, 1993).

Finally, the uncertainty about causation of leukaemia is a reminder of some of the deeper uncertainties about estimates of risks from radiation and about limits to exposure. The estimation of risk factors is inherently difficult when dealing with low levels of radiation because they will only cause small increases in the incidence of disease. The 'ideal' evidence would involve large numbers of people subject to known doses – but where large numbers have been involved (e.g. at Hiroshima and Nagasaki) the doses are not well known and have to be estimated. ICRP moves from its estimates of risk to prescribe dose limits by observing that most people are willing to accept a 1 in 100 000 chance of being killed in a public transport accident, and concluding that the acceptable exposure to radiation is one which gives a 1 in 100 000 risk of dying from leukaemia or cancer. Many people reject this argument and insist that risks should be minimised. This cautious view is strengthened by the growing evidence that risk is greater for some kinds of people (e.g. newborn babies and seafood enthusiasts) and for some kinds of radiation (notably alpha emitters – which were a smaller proportion of the doses from nuclear bombs than they are at Sellafield). These variations in exposure and susceptibility mean that calculations based on average doses and risks may be misleading – but to improve on them requires much more information about actual doses and susceptibility than is currently available.

The fact that the ICRP definition of dose limits incorporates an 'acceptable risk' of cancer of 1 in 100 000, and that those limits were used in deciding the level of authorisation for the sea pipeline, suggests that the notion of an acceptable risk was built in to decisions about emissions. Indeed, BNF do not deny that there is a theoretical risk of a small increase in cancer. Also, Cumbria County Council agreed to increased levels of discharge in 1970 even though they were advised that the increase would cause one additional cancer case per year. As with the thyroid cancers caused by the 1957 fire, such a low increase in incidence would in practice be undetectable and no particular case would be attributable to radioactive discharges.

Even though Sellafield has caused some casualties, it is far safer than heavy cigarette smoking or driving a car (which carry risks of 1 in 1000 and 1 in 10 000 per year). The exact level of risk is not clear, but it is even less clear how an acceptable risk is to be defined. Technical difficulties and questions of values both remain in need of clarification.

The High Court decision has clarified one aspect of the risks of Sellafield: the risks to workers and nearby communities from known discharges are relatively small. This does not satisfy the plant's critics, nor eliminate the possibility of future accidents – though the probability is low. However, recent years have seen major changes in relation to the claimed benefits of the plant. First, the collapse of the Soviet Union has taken away the original purpose – far from needing plutonium for nuclear bombs, the world now has a surplus of plutonium. The prospects of fast breeder reactors becoming technically or economically feasible, and hence providing a demand for plutonium, are remote, and certainly decades away. In the post cold war world, the dangers of proliferation of nuclear weapons or even terrorist use of plutonium make further production of this material undesirable. Second, management of nuclear waste has now become Sellafield's main function: since it receives and stores high-level waste from all Nuclear Electric power stations, this is a function which will go on indefinitely. What is now at issue is whether reprocessing is a necessary part of that waste management. The commissioning of THORP makes it probable that reprocessing will continue at least until German and Japanese contracts are carried out or terminated, though it is by no means certain that this will be economically viable. The operation of THORP will increase radioactive emissions from Sellafield, though probably without increasing doses to the public, and will add to the long-term problem of waste management and decommissioning.

These issues are analysed in *Blunden & Reddish* (1996). It has been proposed (1995) that Sellafield will be the site of BNF's deep repository for intermediate level waste. In conclusion, it is notable that an industry which has existed for only half a century has created a problem of wastes which will remain hazardous to life for millennia. This raises two questions. First, what kinds of decision-making persuade governments to accept long-term risks for short-term benefit? Second, can such wastes be disposed of in ways that will guarantee them against spread in hydrological or geological processes? Even more spectacularly than other environmental problems, radioactive waste directs our attention both to the goals of social change and to the environmental processes that sustain human life, provide resources for the economy and absorb human wastes.

Activity 8

Look back to the two kinds of natural process and four kinds of societal process identified in the summary of Section 2. Do the issues raised in the discussion of Sellafield raise any new aspects of these processes?

4.6   Summary

The discussion of natural, background radiation levels shows that some important natural processes (notably build-up of radon) are far from obvious and also involve actual or potential hazards to human health.

The nuclear industry takes natural materials, including uranium, and processes them to produce substances which do not exist in nature. Some of these, notably plutonium and americium, are extremely toxic and behave in unanticipated ways. Escape of these substances, and others like toxic chemicals, to the environment means that the impacts of advanced technology involve a new degree of threat to health and even to survival.

Production of such hazardous materials may be undertaken for military, political and/or economic reasons and by the state as well as by private corporations. Regulation of such activities has been plagued by secrecy and conflicting interpretations of costs, benefits, and risks.

Pressure groups like CORE and Greenpeace are increasingly appealing to the public at large or to international bodies rather than to national government.

Attempts to develop more rational ways of dealing with hazardous activities include scientific research, better ways of calculating risks, costs, benefits and impacts and the development of new institutions.

5   *Conclusion*

The aim of this chapter was to establish an agenda for the book and series by looking at environmental issues in Cumbria. The summaries and activities at the ends of sections have been designed to help you to think about the variety of, and connection between, factors relevant to environmental issues. The time has now come to summarise these factors and show how they will be addressed.

At the most fundamental level, environmental issues involve an interaction between physical and biological systems and social systems. Three kinds of interaction have been identified:

1   use of natural materials or processes as resources for agriculture or industry;

2   impacts of human activities on natural systems, from clearing of woodland to release of plutonium;

3   natural hazards, of which Cumbria offers only modest examples, such as sudden cold and wet weather in the uplands and perhaps radon in houses and mines.

The history of Cumbria suggests that as society grows technologically more powerful, a process that tends to parallel economic development, the potential size and seriousness of the impacts on natural systems increase. There may be a countervailing trend as better scientific knowledge and new institutions to regulate particular activities and/or the conflicts between activities (e.g. town and country planning, HM Inspectorate of Pollution, the Nuclear Installations Inspectorate) make it possible to anticipate problems and adjust impacts. Very often, however, the economic advantages of exploiting resources or of careless waste disposal override arguments for restraint.

The ability to explain how and why environmental problems occur requires a combination of a scientific understanding of physical and biological processes and also an understanding of the way that society has developed and now operates. For the latter, at least three levels need to be considered:

1   the organisation of production of food and other products (in primitive societies in the past this may have involved simple tools and hand labour to

produce a basic subsistence, but in contemporary society it increasingly involves use of advanced technology to produce for profit);

2   the management of activities or areas by legal or bureaucratic entities, often parts of the state;

3   economic and governmental activities, which are increasingly being monitored and influenced by individuals or groups in a process of environmental politics.

And to understand these levels, you have to recognise the important and complex role of the state.

- It carries central responsibility for regulating activities in its territory, but has other concerns, including military security, economic growth and the balance of advantages between areas and groups.

- It is often involved not only as regulator, but also as producer: for example, as producer of nuclear electricity and plutonium for bombs.

- Increasingly, it is influenced by international pressure as well as by internal interest groups.

The internal and international use of political power and influence is therefore a central issue, both in explaining how things have happened in the past and also in considering how change might be brought about and how individual citizens might act if they are concerned about local or global environmental issues.

Hence natural scientific and social analyses are both essential, but it is necessary to go beyond analysis of how things *have* happened in order to consider how things *ought* to happen, which involves a further step into a philosophical treatment of attitudes and values towards environmental, and indeed social, issues. The study of Cumbria has shown how practical conflicts over railway and waterworks development helped to crystallise new values towards landscape and environment, and how amenity and spiritual values helped to oppose exploitation for profit. Then the movements for national parks and nature reserves developed institutional forms intended to entrench preservationist values. More recently, environmentalist thinking has developed further and has begun to confront arguments that environments have value in their own right and not just in relation to human individuals, groups and societies. These new philosophical values have only just begun to influence practical politics and economics, for example through animal rights groups, but will undoubtedly play a bigger role in future policy debates. The remaining chapters of this book begin to tackle this agenda of issues.

Chapter 2 outlines the history of human impacts, from ten thousand years ago to 1900.

Chapter 3 takes a philosophical viewpoint, identifying a range of environmentalist positions, tracing some of their forebears and clarifying the underlying values.

Chapter 4 discusses the present global problem of development in relation to environment, and systematic ways of approaching this relationship.

Chapters 5, 6 and 7 introduce the scientific view of the biosphere – the geological, atmospheric and biological structures and processes which developed before human intervention and which continue to operate to a greater or lesser degree in spite of human impacts.

Later parts of the series then take up the analysis of productive activities, including agriculture, mining, manufacturing and urbanisation, and of the management and political levels which do much to influence outcomes. At the end of the series, the issues of global change are analysed and all the specific issues brought together in a look at the concept of sustainable development.

References

ALLEN, S. E. *et al.* (1983) *Radionuclides in Terrestrial Ecosystems*, Project 553, Final report to the Department of the Environment, Institute of Terrestrial Ecology.

AUBREY, C. (1993) *THORP: the Whitehall nightmare*, John Carpenter, Oxford.

BLUNDEN, J. & REDDISH, A. (eds) (1996) *Energy, Resources and Environment*, Hodder & Stoughton/The Open University, London (Book Three).

COMARE (Committee on Medical Aspects of Radiation in the Environment) (1988) *Second Report: Investigation of the Possible Increased Incidence of Leukaemia in Young People near the Dounreay Nuclear Establishment, Caithness, Scotland*, HMSO, London.

CORE (Cumbrians Opposed to a Radioactive Environment) (undated) *The Windscale Radioactive Discharges* (pp. 1–4). Available from CORE, 98 Church Street, Barrow-in-Furness, Cumbria.

CRICK, M. J. & LINSLEY, G. S. (1983) *An Assessment of the Radiological Impact of the Windscale Reactor Fire, 1957*, NRPB Report 135, HMSO, London.

EVANS, H. J. (1990) Leukaemia and radiation, *Nature*, **345**, 16–17.

GARDNER, M. J. (1990) Leukaemia and lymphoma among young people near Sellafield nuclear plant, West Cumbria, *British Medical Journal*, **300**, 423–434.

GARDNER, M. J. (1993) Investigation of childhood leukaemia rates around the Sellafield nuclear plant, *International Statistical Review*, **61**, 231–244.

MARSHALL, J. D. & WALTON, J. K. (1981) *The Lake Counties from 1830 to the Mid-Twentieth Century: a study in regional change*, Manchester University Press, Manchester.

MEDVEDEV, Z. A. (1990) *The Legacy of Chernobyl*, Blackwell, Oxford.

MILNE, R. (1993) High Court acquits Sellafield, *New Scientist*, **139(1889)**, p.6.

PEIRSON, D. H. *et al.* (1982) Environmental radioactivity in Cumbria, *Nature*, **300(4)**, November 1982, pp. 27–31.

SUMNER, D. (1987) *Radiation Risks: an evaluation*, Tarragon Press, Glasgow.

SUMNER, D, *et al.* (1991) *Radiation Risks: an evaluation*, 3rd edition, Tarragon Press, Glasgow.

TAYLOR, P. J. (1986) Radionuclides in Cumbria: environmental issues in the international context, in Ineson, P. (ed.) (Institute of Terrestrial Ecology, Grange-over-Sands), *Pollution in Cumbria*, pp. 47–54. Available from HMSO, London.

WYNNE, B. (1992) Misunderstood misunderstanding: social identities and public uptake of science, *Public Understanding of Science*, **1**, 281–304.

Further reading

BINGHAM, H. (1988) *A National Park in the Balance?* GCSE Resource Guide 1, Lake District National Park Visitor Services, Brockhole, Cumbria.

WYATT, J. (1987) *The Lake District National Park*, Webb and Bower, Exeter.

1 Introduction

The purposes of the chapter are as follows:

1 To describe the broad changes brought about by human societies on land surfaces and in the oceans on a world scale from the waning of the ice until the early twentieth century.

2 To give some more detail about these changes for one major region known in part to many of us: the Mediterranean.

3 To discuss some of the explanations given for these changes, noting from time to time that not all interpreters are agreed on the reasons for them, nor indeed on the meaning they may have for society today.

As you read this chapter, look out for answers to the following key questions.
● What were the main impacts of hunter–gatherers and agriculturalists on the environment?
● How did industrialisation transform those impacts?
● How have human impacts developed in the Mediterranean region?
● Why is it difficult to explain these changing impacts?

As we saw in Chapter 1, northern Europe was covered by ice sheets during a succession of glacial periods over the last 2 million years. The last glaciation ended about 12 000 years before the present (BP). Since then, the melting of the ice has made available an increased area of land surface for human occupation. The continents and the seas assumed their present form and distributions by about 9000 BP. By 10 000 BP, the human species had spread from beginnings in Africa some 2 million years earlier into all those parts of the globe able to support human life at that time.

Before 10 000 BP virtually all humans lived as food collectors (usually known as **hunter–gatherers**) rather than food producers. Their success in colonising an Earth which was biologically and climatically diverse demonstrates their ability to subsist in different environments. The human alimentary system can cope with many kinds of food. Thus, where plant material comprised the bulk of available food, humans could be largely vegetarians (as with the Bushmen of the Kalahari Desert), and where vegetable food was scarce but animals were abundant they could be largely carnivorous, as with the Inuit (Eskimo) of the Arctic.

In an atlas or textbook, a world map of 'natural' vegetation is a map either of what that vegetation would be like if human influences were removed and natural processes took over, or vegetation as it was before human influence was as strong as it is now. In effect, the map is that of the vegetation as it approached a kind of equilibrium after the major climatic changes that followed the melting of much of the ice, a process completed at varying rates but mostly in place by c.9500 BP. The pattern established (see Chapter 6) is a familiar one: the bands of vegetation form a series of latitudinal zones, with the tundra near the poles and tropical forests at the equator. This orderly sequence may be interrupted where there are mountains or areas of especially low rainfall. Each of these biomes has its characteristic animal populations and plant life and each presents to the hunter–gatherer a set of opportunities as well as constraints. Shallow seas may be especially fruitful in their year-round yields of fish, for example, whereas migrating deer may only be available to hunters for a short season. In tropical forests much of the animal life may be at tree canopy level rather than on the ground, and so a specialised technology must be developed to kill animals such as monkeys: the blowpipe and poisoned dart is one example. So closely were hunter–gatherers tuned to these various environments that they have traditionally been regarded as inhabiting a kind of paradise in which they were just one component of the complex ecological web. In contrast, this chapter will show that hunter–gatherers had significant impacts on natural environments, though their impacts were less than those of later agricultural and industrial societies.

The world has changed a great deal since 10 000 BP and much of that change has been at human hands. The complexity of those alterations is immense, so to avoid getting lost in a welter of detail, some kind of generalising framework for the course of history is needed. A fuller treatment, such as that in Simmons (1993), has to take account not only of human biology and the physical environment, but also of the ways in which varied human groups use technology – that is, the process of using natural resources to achieve human purposes, such as food, shelter, fuel and manufactured goods. So, the assumption is made in this chapter that the ability to get a living from our natural surroundings and to change them is influenced by the success of humans in getting access to energy. That is not to say that energy in its various forms *determines* all human **culture**, because this, whether that of a hunter–gatherer or of a modern society, also brings together population, resources, technology and social rules, values and beliefs. Nevertheless, the possession of an energy surplus beyond what is needed to survive opens the way to bring about all kinds of change, mostly through the medium of technology.

Humans use energy in two ways. The first is the use of energy derived from food to maintain life and to work directly on their surroundings. This is called **somatic energy**. The second is the harnessing of energy sources outside their bodies, e.g. that of animals, wind, water and fossil fuels, using a variety of technological devices. The latter, **extra-somatic energy**, multiplies our ability to get access to resources and to change our surroundings.

So the quantity of energy to which we have access, both somatic and extra-somatic, and also the source of this energy, can provide the framework we seek to give form to the course of human history in its environmental context.

Figure 2.1 shows the great increase in energy used by food-collectors, agricultural and industrial societies, as well as indicating some variation within these broad categories.

Q   In Figure 2.1 how much more energy is used per person in advanced agricultural society and in technological society than by hunters?

A   Advanced agricultural societies used five times as much energy per person as hunters and technological societies nearly fifty times as much. In fact these figures are variable between places as well as between stages and some technological societies use energy more efficiently than others, i.e. they use less energy for the same effect.

Although hunter–gatherers lived at low densities and possessed only simple tools like spears, bows and digging sticks, their harvest of animals and vegetable products could still significantly affect the balance of natural ecosystems. These effects became more dramatic when some groups gained control of fire, since this could greatly alter vegetation cover.

Further changes occurred when some groups settled in one area and began to practise **agriculture**. This depended on the ability to domesticate animals and plants. Much more energy had to be expended in clearing, tilling and planting land, but the rewards were a much higher output of food per unit area and hence the ability to sustain much denser populations. Control of water for irrigation represents an extreme case of high inputs to secure high outputs, with consequent transformation of natural ecosystems. In some groups, domesticated animals were used to supplement human energy, e.g. for transport.

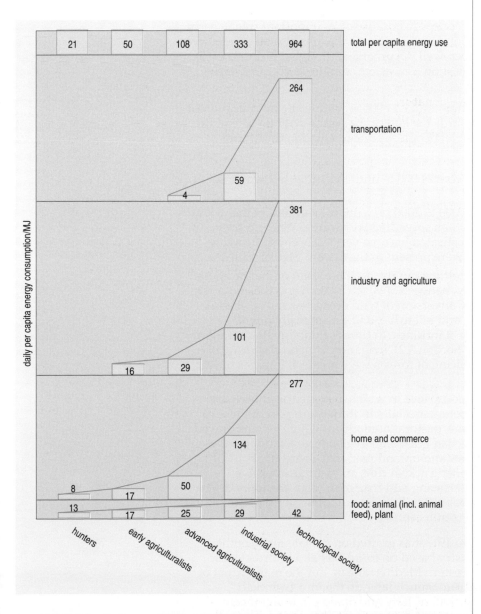

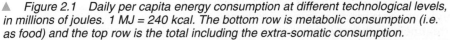

▲ *Figure 2.1   Daily per capita energy consumption at different technological levels, in millions of joules. 1 MJ = 240 kcal. The bottom row is metabolic consumption (i.e. as food) and the top row is the total including the extra-somatic consumption.*

The industrial revolution was associated with the use of fossil-fuel energy (see Section 4.1) as well as with the creation of new machines and forms of social organisation. Progressively, such innovations multiplied humanity's ability to alter natural environments, both positively to produce an abundance of new products and negatively through loss of amenity and new forms of pollution. The impacts of technological society are central to later chapters, but this chapter concentrates on describing and analysing the changes made by a variety of human societies up to the early twentieth century.

Although examples will be drawn from the whole world, a better understanding of the chronology of social and technical change can be gained by a more sustained focus on a single region. The Mediterranean basin has been chosen for this purpose because of its important role in early agriculture and in the evolution of 'western' culture. It is also an area where environmental change has been well documented, though it is not so easily explained. The Mediterranean shows evidence of all three cultural stages, but especially of agriculture:

1   Hunter–gatherer occupation until *c.*11 000 BP, with some modification of the landscape happening towards the end of the period with the use of fire by early Stone Age (Palaeolithic) people, as documented in the Mount Carmel caves.

2   Agricultural occupation from 11 000 BP until the 1920s, but with a set of subphases:

(a) Early agriculture from 11 000 to 7000 BP, with its heartland in the mountains of the Levant and then spreading westwards. Wheat, barley, goats and dogs were all domesticated here, so very early cereal-growing and stock-raising economies were present in the eastern Mediterranean.

(b) the consolidation of agriculture and the establishment of the traditional Mediterranean rural economy, between 5000 BC (7000 BP) and the end of the Byzantine Empire in AD 640. Conversion of wild ecosystems to agriculture and pasture was widespread and intensive, and the intensity often fluctuated according to political fortunes. In times of stability, the creation of terraces on steep land (see Section 3.2), irrigation, upland pasturing of animals, lumbering and coppicing of forests were greatest. In times of war, forests were often burnt.

(c) A period of agricultural occupance in which evidence of soil erosion and desiccation is more frequent, especially in the eastern part of the area. In areas conquered by Muslims, pastoral nomadism (where groups moved to new areas as soon as flocks had grazed all available vegetation) was ever more frequent, affecting forests and upland pastures. Some areas experienced depopulation under Turkish rule, and labour-intensive agricultural systems such as terracing and irrigation were abandoned. The population of Israel was probably 2 million in early Roman times but only 300 000 at the end of the nineteenth century.

3   This phase, from AD 640 to 1918, has grafted on it a late nineteenth-century development of industry based on fossil-fuel and hydro-electric power in a few parts of the western Mediterranean, but the main thrust of modernisation of economies came much later, in the mid-twentieth century and thereafter. Nevertheless, Italy and Spain made significant strides towards becoming developed economies in this period. The full flourishing of the industrial and nuclear periods in this region is not, however, afforded extended treatment in this chapter.

Activity 1

Reflect on the areas familiar to you, because you live there or have taken holidays there. Are there any examples of natural, as opposed to agricultural, ecosystems surviving today? What impacts have past societies had and how has industrial society altered agriculture? Draw up a list of your current beliefs and reconsider it when you reach the end of the chapter.

2   Hunter–gatherers and their environments

2.1   Introduction

In this section we review briefly the variety of natural environments that have been occupied by hunter–gatherers during the last 12 000 years or so (known to earth scientists as the Holocene) and also look at the energy flows which linked the people to those surroundings and the ability of some of them to control this energy and manipulate the ecological systems in which they found themselves. The evidence comes from archaeological sources for the earlier part of the period and from work such as ethnographic surveys (studies of surviving tribes) for more recent times. The number of hunter–gatherers declined as communities settled, as agriculture became intensified, and **industrialisation** took over in larger communities. Twelve thousand years ago the entire human population consisted of hunter–gatherers. Now perhaps 0.001% follow that way of life, and their lives have been influenced by the higher energy-intensive world around them (e.g. the Inuit now use snowmobiles for travelling).

The link between hunter–gatherer societies and Nature is through technology, i.e. the exploitation of existing knowledge of materials and energy, for productive ends. This knowledge has changed and developed over time. At the end of the Pleistocene (the geological epoch that preceded the Holocene), hunter–gatherers had access only to organic materials and to stone, but thereafter they quickly made use of metal when it became available. In more recent times, relict hunter groups have rarely been reluctant to absorb the products of the industrial revolution when these came their way. The traditional technology centred around wooden tools for grubbing up plants, baskets or slings in which to bring plant materials back to camp, hunting aids such as the spear, bow and arrow, blow-pipe, slingstick, woven nets, and poisons. More recently the rifle, the outboard motor and the snowmobile have slipped into this repertoire with ease.

Q   Figure 2.2 shows a map of near-recent hunter–gatherer groups. Given the variety of climatic zones in which these groups persist, can you suggest one similarity between them?

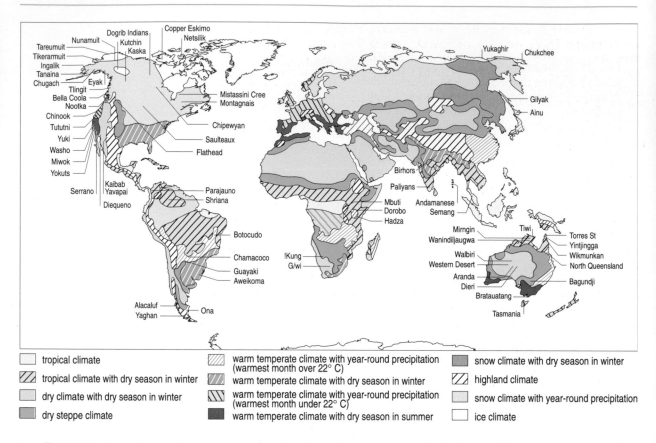

▲   *Figure 2.2   A world map of some near-recent hunter–gatherer groups. In total, these will comprise a very small percentage (probably less than 0.001%) of the world population.*

A   They are all extreme climates, whether cold, hot, dry or wet, and hence 'marginal' in terms of conventional agricultural use.

Before the nineteenth century hunter–gatherers interacted with Nature by their collection of plant and animal material within the territory of a particular group. In areas where food changed according to seasonal availability, and often also because of kinship and other social arrangements, it was desirable or even essential for groups to move in a regular annual pattern round a territory where they had rights of access to the resources. In areas where resources were not scarce, hunter–gatherers who maintained relatively low population levels could simply live off the **usufruct** of Nature, i.e. taking a small enough proportion of natural production to leave the ecosystem unchanged. For example, a 1960s study of the !Kung bushmen of the Kalahari found they were dependent upon the nut of the mongongo tree as a staple food even though their total dietary spectrum was very wide. But they had no need to plant mongongo trees to ensure their pollination or any such manipulations, because enough nuts could be gathered in season to feed everybody with a decent diet of calories and protein. Their neighbours, the G/wi bushmen, relied on the *tsama* melon for part of their diet and its abundance was variable: thus the G/wi had to travel further in a bad melon year to get their required quantity.

This rather Romantic image of children of Nature did not hold everywhere. There were environments where the people felt that Nature would not always provide unless some action was taken by them. For some this was mainly of a non-material kind in the shape of practices designed to propitiate the gods into maintaining food supplies. For others it was a case of the gods helping those who helped themselves and so management of plant and animal populations was practised.

In the case of plants, this might mean occasionally diverting some water over a stand of wild grasses, for example, to ensure their seed crop was heavy, or perhaps transplanting some wild yams into an easily accessible location or one inaccessible to natural predators. Animal populations might be protected from hunting in some areas so that depleted numbers might expand again (a variety of social mechanisms might exist for this purpose), or there might be a prohibition on the killing of pregnant females. In semi-arid environments such as parts of Australia, canals might be dug in and near swamps to maintain the right habitat for eels.

The methods of achieving management were mostly through systems of social organisation. Most such societies have a complex set of rules as to which resources may be used, when, and by whom, often with a mythical structure and oral tradition that reinforces the appropriate practices. In many societies there were duties to supply the young and the old. There was some exchange between peoples over longer distances of, for example, stone tools or salt, but control over rates of use of environmental resources remained with the local inhabitants, who would be directly affected by the consequences of unwise action.

On a world scale, there was a broad gradient of different use of natural resources for food, e.g. hunting of animals in high latitudes, fishing in mid-latitudes, and the gathering of plant food in the tropics. But in all groups hunting seemed to provide at least 20% of the diet. Plants must therefore have been key foods in all except extreme environments like the Arctic. The hunter–gatherer way of life seems to have been founded upon the consumption of as much meat and fish as possible and as much plant food as necessary.

2.2   Ecological change produced by hunter–gatherers

As discussed above, not all hunter–gatherers changed the ecology of their environment, but some did so in both temporary and permanent ways. Impermanent alterations often centred upon the impact of a hunting group upon an animal population. This could be severe but yet not so damaging to the reproductive capacity of the beasts that their numbers could not recover. Buffalo hunting on the high plains of North America before the advent of Europeans is one example. Numerous archaeological excavations have shown that herds of bison (called buffalo in North America) were driven over cliffs, into box-canyons or into dune slacks and then slaughtered wholesale. Yet the animals' availability was not reduced in the long term because in general they concentrated on herds of males and so let the females carry their young unmolested. Analogous practices were carried on further north, where native American people in the boreal forest developed traditions of resting particular areas within their territories so that, for example, elk (moose) and beaver populations might recover from heavy hunting. The occasional use of fire as a hunting aid (i.e. to drive out animals from cover rather than to try to change the vegetation) might well

not change the ecology for more than a season or two; recovery of the vegetation to the original structure would have been likely even if the species composition was in some way altered.

Of more interest and significance, however, are the cases where hunter–gatherers have affected their environments permanently. One of the most discussed issues is the rapid rate of extinction of the megafauna which had existed immediately before the arrival of humans. These extinctions were most rapid in North America and Australia. The phenomenon is perhaps most closely observable in North America: here, two-thirds of the large mammal fauna present near the end of the Pleistocene about 15 000 years ago disappeared, including 3 types of elephants, 6 types of giant edentates (armadillos, ant-eaters, pangolins and sloths), 15 types of ungulates, and various giant rodents and carnivores. There is no evidence of such extinctions in earlier periods which might have been expected if climatic changes in the Pleistocene had been the cause, nor any firm evidence of survival of these types of mammal past the 500-year period during which the extinctions appear to have taken place. These data have led to the hypothesis that the cause was the sudden influx of humans and their predatory habits, against which these mammals had no genetically implanted defence behaviour. The date that humans crossed from Asia to North America via the Bering Straits land bridge (iced over in the last glaciation) is usually taken as being about 12 000 BP, and a simulation of the extinction pattern supposes a wave-front advance by hunters whose

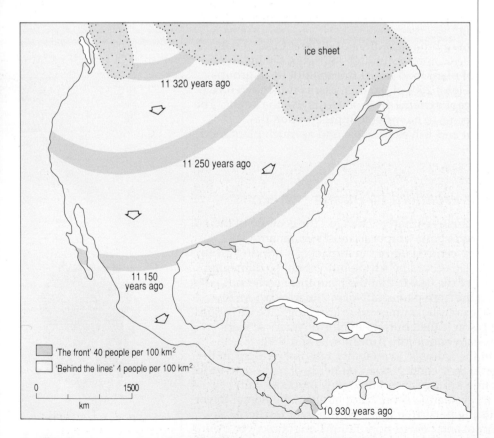

▲   *Figure 2.3   A hypothesis for 'Pleistocene overkill' sweeping through the North American continent, with the initial colonisation occurring from the north.*

population periodically exploded because of the new and favourable habitat (Figure 2.3). At the advancing front, a human population growth rate of 2.4% per year may have given a population density of 0.4 persons per km^2 and a forward movement (due to hunting-out of the large mammals) of 16 km per year. Thus the front could sweep from Canada to Mexico in 350 years and extirpate the large mammals through superior predation techniques. Some support is given to this 'overkill' hypothesis by the effects of the arrival of humans into other hitherto unpeopled places. The disappearance of the moa bird from New Zealand occurred within a few hundred years of the human occupance of the islands; the megafauna of Madagascar (including a large terrestrial bird *Aepyornis*, and a pygmy hippopotamus) disappeared within a similar period after the first human occupation in AD 1000; in Java and Sulawesi, populations of dwarf elephants likewise did not long survive the coming of humans. It would appear, therefore, that when introduced in a new habitat man the hunter is capable of rapidly exterminating flightless birds and large mammals.

Q   What was the crucial difference between the early hunting of the megafauna and the later hunting of North American bison?

A   The megafauna was hunted by newly arrived and expanding human populations. The native Americans in the prairies had been in an equilibrium with their environments for hundreds of years.

More recent permanent changes by hunter–gatherers can be studied more easily. The management of fire became an important skill and its regular use created permanent change. The aboriginal inhabitants of Australia provide an example. In the interior, fire was regularly used as a hunting aid since it flushed many animals out of the bush and from underground, and plants that survived were those able to regenerate quickly in such circumstances. The result was that the vegetation types encountered by the newly arrived Europeans were in fact fire-adapted ecosystems produced by human agency. In the north, women regularly fired the vegetation

◀   *Australian eucalyptus trees of the same age, having germinated after forest fire.*

containing a particular cycad tree: it then produced more fruit and also produced it almost immediately, which is always an advantage to people on the move. In North America, the native populations of the forest–grassland edge also used fire as an aid to hunting, including keeping down the quantity of undershrubs that prevented a hunter getting a clear sight-line for his arrows. This habit maintained a mosaic of open woodland and grassy glades. When the native Americans were extirpated, the forest rapidly reclaimed the land.

We have an instance of permanent change from the uplands of England and Wales during the period of the last hunter–gatherers, known to archaeologists as the later Mesolithic of *c*.8000 to 5500 BP. The present-day moorlands such as the Pennines, Dartmoor and the North York Moors have yielded palaeoecological evidence (from soil samples, pollen analysis, etc.) to show that in those times such uplands were largely deciduous woodland. Yet among the woodlands there were clearings which seem to have been maintained by fire, and their frequency in time and space is such that natural causes are unlikely. In some places, these openings disappeared when agriculture started, which suggests that burning was an integral part of the hunter–gatherers' way of life. But on other sites, the removal of the trees allowed the soils to become waterlogged (because of reduced evapotranspiration, discussed in Chapter 5, Section 5.2) and peat grew (Figure 2.4). Over time, heavy rain percolating through acid peat dissolved plant nutrients from the soil below, leaving a grey-white impoverished sandy soil called a podzol, on which only heather and coniferous trees will grow. In waterlogged places, the peat grew to depths of 2–3 metres and formed a blanket over the land. This condition, known as blanket bog or blanket mire, can still be seen today (though often now eroding) and is an example of a landscape element formed by hunter–gatherers (Plate 11). See Chapter 6 for more on such ecosystem interactions.

Further examples of more permanent change can often be seen when hunter–gatherers came into contact with agricultural or even industrial populations. For instance, many areas of the boreal forest were almost entirely depleted of fur-bearing animals by indigenous trappers, who sold

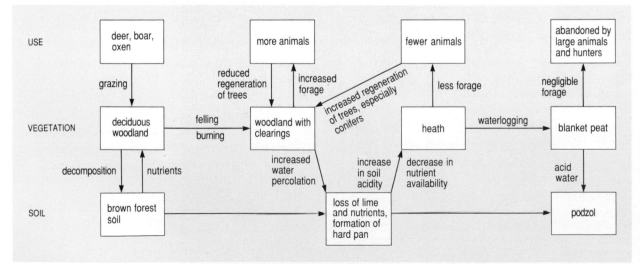

▲ *Figure 2.4   A sequence of possible processes in the clearance of forest in upland Britain by hunter–gatherers, ending with either a mosaic of woodland and open vegetation or blanket bog.*

their catch to the Hudson's Bay Company and similar agencies. The insatiable European market for beaver, for example, wiped out many populations in spite of the Company's efforts to install rotational trapping schemes. Had fashions not changed, the beaver might have disappeared from most of Canada, as it did from many areas of the USA. The balance between hunter–gatherer societies and their environments was often disrupted by trade, disease or missionaries even before agricultural settlers moved in to displace them.

2.3   Mediterranean hunter–gatherers

We have rather patchy information about the economy and landscape relations of hunter–gatherers in the Mediterranean before the development early in the Holocene of agriculture in its various forms. During the last glaciations of northern Europe, the predominant vegetation in the Mediterranean seems to have been an unforested plain (steppe) dominated by wormwoods and goosefoots (*Artemisia* spp. and Chenopodiaceae), with some oak woodlands. When the climate improved, there came more oak woodland, with juniper and pistachio admixtures. This woodland seems to have been open, with glades where many wild grasses grew, some of which were ancestral to cultivated cereals. In these surroundings, hunter–gatherers lived off many animal species, including the onager (a wild ass), wild ox, red deer, wild goat and gazelle. Among the important plant species was undoubtedly the oak, whose acorns were ground into a flour. Some flint blades belonging to that era have been found which show evidence of wear from harvesting of wild grasses, though whether as fodder for early domesticated animals or for human food is not known. Early Holocene sites in Palestine have yielded evidence of a concentration on the gazelle as a source of meat, leading to speculation that attempts at domestication had been made. Conclusive evidence of permanent environmental manipulation by humans in the late Palaeolithic is sparse, but several scholars have extrapolated from the evidence of hearths in caves such as those on Mount Carmel that these people controlled and used fire and probably caused landscape change. They point to the advantages of fire to hunters; for example, animals are attracted to burned-over patches of ground since these often sprout plants quickly after fire, even before the autumn rains. Such areas, too, often grow grasses, bulbs and tuberous plants which add to the food supply of human groups. So it seems probable that the long association of fire and humans in the Mediterranean started in pre-agricultural times. The region is of course especially well suited to this relationship because of the long and dry summer which provides in most vegetation types a stock of fuel that can be ignited by lightning or by human agency.

Hunting continued even when agriculture was established as the dominant way of life, as it still does. Some hunting was for the pot and some simply for pleasure. In the Mediterranean, an historic form of hunting that had a strong impact upon ecosystems was the procurement of live animals for the Roman circuses. Since they had to be seen by thousands of spectators at once, only larger species were captured: elephants, ostriches, lions, leopards, hippos, tigers and crocodiles all featured at one time on the programme. With his talent for the unusual, Nero once put on a show of polar bears catching seals. At the dedication of the Colosseum, 9000 animals were destroyed in 10 days; Trajan's conquest of Dacia was celebrated with the slaughter of 11 000 wild animals. When it is remembered that these are often large and long-lived herbivores or predators from the tops of food chains, then it is scarcely surprising to learn that the elephant, rhino and zebra were extirpated in North Africa, and lions in Thessaly, Syria and Asia Minor. This process would have interacted with land use change and the expansion of pastoralism, both inimical to the presence of wild animals, especially predators. Thus, it can be surmised that many inhabitants from those regions, as well as the enthusiastic spectators, would not have objected, in much the same way that people still call for a renewed onslaught on the wolf in Spain and Norway today.

Activity 2

Review the section on hunter–gatherers and consider how their impacts on natural ecosystems differed from the impacts of a similar-sized omnivorous, non-human animal.

2.4   Summary

Early hunter–gatherers living off the usufruct of Nature must have had little impact on their environment. More sophisticated weapons and control of fire increased their ability to change the proportions of plant and animal species, whether accidentally or in deliberate attempts to increase the availability of food.

These groups are characterised by a range of social practices, as well as technological ability, including rituals intended to propitiate the spirit world, rights over and duties towards other people, and in many cases rules governing the exploitation of the environment.

A culture, whether that of a hunter–gatherer or of a modern society, brings together population, resources, technology and social rules, values and beliefs.

The long-term sustainability of hunter–gatherer groups must have been contingent on cultures which did not degrade the productivity of the ecosystems on which they relied.

3   Agriculture

3.1   Origins

This section deals with the period between about 8000 BC and AD 1800 when agriculture replaced hunting and gathering as the principal mode of food-getting but was still basically dependent on energy from sunlight.
The next section will deal, among other things, with agriculture in an industrial era.

Compared with hunter–gathering, agriculture represents an intensification of use: more output is achieved per unit area per unit time

◄ *Egyptian tomb painting showing early agriculture.*

than was possible with hunting and gathering. This is done by using plants and animals culturally selected for cropping. The end product, after selecting the best species or breed for the purpose over generations, is a **domesticated** plant or animal with a life cycle that has been so tailored to meet human demands for food, fibre or skins that it has lost its ability to survive without human help. Pre-industrial agriculture had to expend a lot of energy in tasks such as sowing, ploughing, fencing and herding, but the system could only persist if a positive energy balance was maintained by yielding more energy than was invested. Yet at its most productive, agriculture provided magnificent surpluses: enough to feed those who built the pyramids of Egypt, for example, or to support the kind of cities in which Mozart lived.

The beginnings of agriculture are still only partly known. A number of places in the world were important in seeing the emergence of different agricultural systems but there are some domesticates whose origins are not known. In the hill-lands of western Asia, cereal-growing around permanent villages grew up in the period either side of 7000 BC, and these people kept domesticated animals as well. Perhaps nomadic pastoralism, based on herds of domesticated animals, became fully established about 4000 BC. In south-east Asia, rice was domesticated around the same time as the western Asian cereals like wheat and barley; millet arose from northern China in the same era. In the period between 4000 and 2000 BC, New World agriculture arose on the basis of maize, potato, beans and squashes. From these origins, the various types of agricultural system spread and developed into most parts of the world. They often replaced hunting and gathering on the way, though were unable to oust the traditional system from the more marginal places such as the very dry, the very cold, and the remotest tropical forests. Such was the success of agriculture that the human population grew from perhaps 170 million in AD 1 to 900 million in AD 1800. By that time some agricultural areas had moved far beyond production for subsistence and involved many economic transactions with the non-agricultural sector, as shown in Figure 2.5.

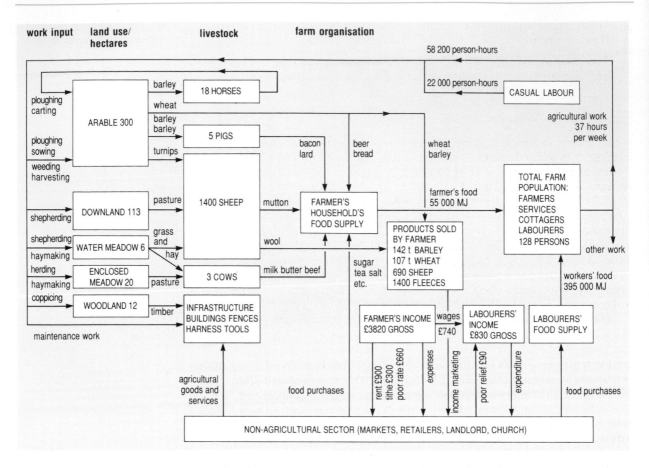

▲   *Figure 2.5   The energy and material flows of a pre-industrial farm in Wiltshire, England in 1826. This shows the energy needed from humans and animals to maintain a productive system with surpluses to sell off 107 tons of wheat, 142 tons of barley etc. In a truly subsistence system the whole system would be self-contained. Note that here there is no import of commercial energy into the system.*

Activity 3

Imagine you were the farmer described in Figure 2.5 and draw up a rough balance-sheet of income and expenditure. You will find that the diagram specifies some items by number, some by value and some by both. Compare your answer with that supplied at the end of this chapter, and then answer the following questions.

Q   What do these transactions imply about the culture in which this farm existed?

A   Not only were there well-developed markets for agricultural produce, labour and services, but the existence of landlords, farmers, labourers, shopkeepers and clergy implies that there was a complex social system. Britain was by then a leading trading nation and was becoming a colonial power. Many factors apart from environment were influencing the way land was used.

Q   What important product of animals and people is not identified in this
    flow diagram?

A   The manure, which would be a vital means of sustaining soil condition
    and continued productivity.

3.2   Global impacts

If agriculture represents a concentration of energy input by human cultures
leading to increased intensity of yield, then the scope for enhanced
environmental impacts is correspondingly higher. Conversion of natural
systems is deliberately undertaken and on a much larger scale (both
spatially and ecologically) by agriculturalists than by hunters. One of the
greatest changes wrought by humans has been to clear woodlands
permanently in order to grow crops. Examples abound: the deciduous
forests of Europe yielded to the plough in prehistoric and medieval times;
those of North America disappeared along with many of their indigenous
inhabitants after the European colonisation. Analogous stories can be told
from Asia, the former USSR, South America and Oceania and the process is
still continuing, especially in tropical rainforest areas.
   Clearing a forest has serious ecological consequences: soils erode more
easily and water is shed more quickly so that floods, silting up of streams
and lakes, and dumping of nutrients are all common downstream from the
deforested area. To combat loss of soil on sloping land (especially in areas
of seasonally intense rainfall), many societies constructed terraces, which
are series of flat areas on different levels, sometimes with the soil retained
by a wall. In essence, these convert a sloping field into a flat one, so that
both soil and water are held up for further use. Water is often seasonally
scarce and so small reservoirs are built to store it. Terracing in south-east
Asia converts slopes and flatlands alike into aquariums in which *padi*
(paddy) rice and also fish and shrimps are grown for food, and multi-
cropping of the cereal is often possible. All these systems must mimic the
natural condition in circulating nutrients and so humans must fertilise their
crops with animal manure, night-soil, silt, soot, marl or whatever else is to
hand.
   Such irrigation systems imply complex forms of social organisation as
well: in Bali today the temples are the main nodes for the organisation of
the water distribution. In ancient Mesopotamia and Egypt it seems likely
that authoritarian rule was necessary to keep the irrigation systems in
working order: they were so large-scale that only centralised decision-
making could have kept them viable. Thus we recognise that politics (i.e.
the exercise of power) is an integral part of the way in which an
environment is used. An imperial power, for example, can reach out into its
empire for resources and effect environmental changes which would not
otherwise have occurred. In the Mediterranean the Roman Empire dragged
in wild animals for its circuses, and also caused many areas to be planted to
wheat. The British Empire set up rubber plantations in Malaya, for instance,
and tried to do the same in post-1945 Tanganyika for ground-nuts but met
resounding failure for largely environmental reasons. Today, deforestation
in Amazonia or in south-east Asia results from international trade in timber
and meat (cleared land is used for cattle ranching) as well as from the needs
of the populations of those areas. The empires are now economic rather
than strictly political.
   Less intensive forms of cropping produce their own levels of impact.
Shifting agriculture (where forest is cleared and crops cultivated for a few

years until fertility declines) allows the original type of ecosystem to return, but often the actual species composition is changed: the more frequent the clearance, the greater the change. Nomadic pastoralism also changes the composition of vegetation. Domestic animals are selective in their feeding habits and so certain species are preferentially eaten. Above some threshold level of consumption they will not regenerate, leaving the site to thorny or toxic plants or perhaps simply to bare soil, leading to the risk of desertification. Many high mountain pastures in the Andes are the result of centuries of grazing as much as of climate and soil. The same can be said of the grassy areas of the moors of Britain: the species composition reflects the density of sheep, with the wiry *Nardus* predominating where grazing intensity is high.

Agriculture (including pastoralism) is central because it feeds people. Thus most societies make attempts to extend its area, either to feed more people or to feed some of them better. But pre-industrial societies also reached into their environments for a variety of other purposes: for materials to use in construction and as fuel, for example. They also sought pleasure in parks and gardens and, like most humans before and since, engaged in warfare which was sometimes at the expense of the environment as well as of themselves.

Extending the agricultural area often meant land transformation, usually called reclamation. Areas already relatively flat, such as salt-marshes and other estuarine areas, have a long history of reclamation, as has the draining of wetlands. In the former, the ability to build a good sea-wall has always been crucial and in the latter the windmill was an important piece of technology in using the energy of the wind to lift water out of one channel into another to speed it to the sea. Many other wild ecosystems have also been reclaimed. Heaths and moors in Europe, for

▲ *Early societies dug peat for fuel; the Norfolk Broads were formed when these areas were flooded.*

example, were targets in the medieval period. The Cistercian monks were especially active in those regions because they were supposed to live away from the temptations of the world, and what better place than a desolate moorland valley? Forests have always been useful for their products: wild foods to supplement basic diets are usually present and there is above all wood. Most pre-industrial societies depend upon wood for all their fuel requirements (domestic and industrial), for much construction (shipbuilding, scaffolding, the frames of buildings), for animal fodder and for a host of small things like tools. Woods can be managed to supply these materials: for example, coppicing and pollarding provide a supply of poles (for fencing or for charcoal-making); shredding (i.e. stripping branches) is a good source of animal fodder. Deciduous trees growing in semi-open conditions may develop crooked branches which are essential for ship timbers; in contrast, dense stands of conifers produce the straightest masts.

The mention of industry reminds us that this was not exclusively a feature of the nineteenth century and after: there were workshops and small factories, quarries and mills in many parts of the world by 1800. Some needed power and this was provided by the diversion of stream courses into watermills; others needed heat for smelting, for example, and so woods might be managed for a continuous supply of charcoal, like those of the Sussex Weald in the fifteenth century. Wastes began to be a problem: the tanning industry has always produced noxious effluents and in pre-industrial Holland special drains (*stinkerds*) were constructed for tanning wastes.

Where agriculture produced good surpluses, a leisured class might be sustained. This might devote itself to religion, to learning, to building monuments to itself, to acquiring an empire, or simply to pleasure – at any rate some of the time. Thus we have environmental manipulation for non-material purposes. For instance, most gardens produce useful things to eat, and often herbs and medicines, but even so there is always an accent on pleasure. In many cultures, combinations of shade trees, flowers, water and grass (and moss in Japan) reflect particular values of that society: the desert origins of Islam, for example, are shown in the primacy of water in their gardens. Other values may be shown, as in the Gardens of Love in the courtly era of European history. Killing of wild or semi-wild animals for pleasure appealed in most cultures, and to ensure the day's success walled parks were often constructed (in places as far apart as England and China) to confine the quarry: deer were very popular, as were wild boar. The opposite might also apply, with edicts for forest preservation (under the Buddhist rulers of India, for example) and prohibitions on killing certain species, like elephants in India in the third century BC or kingfishers in China in 1107.

Q   So far, this account has stressed the effects of more sophisticated agricultural technology, larger scale markets and political units. Does this imply that agricultural cultures are more rational (believe in reasoned thought rather than divine intervention) than hunter–gathering?

A   Not at all: this period has also seen the origin and spread of major religions. Of these, Christianity has been associated with successful political and economic empires. Although it is open to many interpretations, many Christians have interpreted scripture as authorising humanity to multiply and to assert domination over Nature. Later and more diverse attitudes are discussed in Chapter 3.

Another limitation of human rationality is that throughout history wars have been common and the environment has sometimes suffered: at the battle of Pylos all the vegetation of one island was burned off so that the Athenians could see the movements of the Spartans; similar depredations were made on Scottish forests during clan conflicts. The Romans sowed the fields around Carthage with salt and wells were often poisoned. Most of these changes were temporary, however. At Massalia, Plutarch put the positive case: so many Teutons were killed that the soil 'grew so rich and became so full to its depths of the putrefied matter that sank into it that it produced an exceeding great harvest in after years'.

3.3   The agricultural period of the Mediterranean Basin

> Even places alter . . .
> . . . where the sea once spread, on steady land
> Now houses, trees and men securely stand . . .
> Torrents a valley of the plain have made
> And mountains headlong to the sea conveyed.

So wrote Ovid, though more elegantly than this translation suggests. Nevertheless, a certain climatic unity, a long and varied cultural history and relative familiarity all make this region a suitable place for the study of environmental change at human hands during the agricultural period of human history.

The basis of the traditional agricultural system of the Mediterranean region was wheat, supplemented by a variety of tree and bush crops. Of these, olives (usually pressed to oil) were an important part of the diet since they added to the fat content, as did a number of nut crops. Fruit was important, and the vine supplied carbohydrate as well as a highly tradeable commodity and a demotic source of pleasure. This historic mix was much

▲   *Vineyards in southern France.*

diversified by the Muslims, who brought sugar, citrus fruits and rice, as well as the carnation and the rose. The place of animals in Mediterranean agriculture is sometimes underestimated: they were a valued component of the diet and were if possible pastured near the village. If no forage was available, then the flocks might be taken on long journeys to areas of summer pasture in mountain areas (the process called **transhumance**): from Provence to the Alps of Savoie, for example. Transhumance spreads the environmental impact of grazing, which in the Mediterranean most often involved sheep and goats rather than cattle, which have high daily water needs. Only in a few places could irrigated land be spared to grow fodder such as lucerne.

The vegetation type we now associate most with the region is the low scrub of oak and juniper, interspersed with grassland or with tufts of flowers. This is the product of fire upon the former forests, and is called the *maquis* or *garrigue*, *phrygana* in Greek, and *tomillares* in Spanish (Figure 2.6). Two thousand years ago, Virgil recorded 'scattered fires, set by the shepherds in the woods, when the wind is right', and the Arabs carried on this practice for at least another thousand years. Once created, the *maquis* is very dense unless subjected to more fire, and so the shepherds carried on burning to produce the vegetation community they wanted. Where frequency of fire and density of animals has been high, the result is a low sward of asphodels, brilliant in spring but soon burned off by the sun.

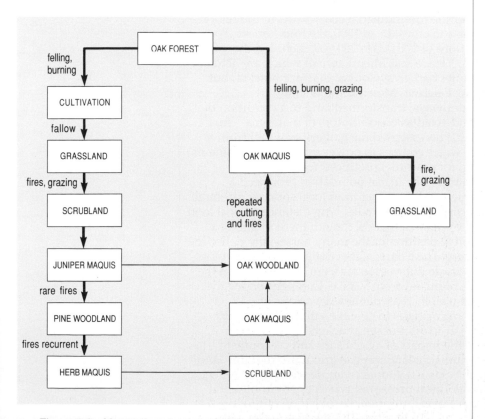

▲ *Figure 2.6   Vegetation changes under different land use regimes on limestone uplands in Mediterranean lands. The starting point is the 'climax' vegetation of evergreen oak* Quercus ilex *and the thick lines are processes of active management by human groups. The thinner lines are succession after relief from those pressures: note that recolonisation to oak forest is possible.*

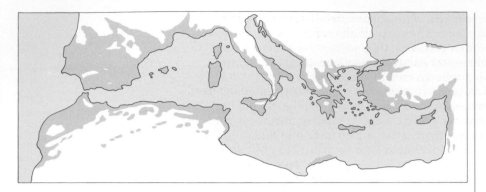

▲ *Figure 2.7   The distribution of shrubland vegetation of the* maquis *type around the Mediterranean basin. This is a generalised distribution: your glass of retsina comes from a vineyard, not a scrubland.*

Today, though, this vegetation type is widely distributed around the Mediterranean, as shown in Figure 2.7, and accounts for 15% of the land cover of Greece, for example, with some 10 000 hectares per year still subject to fire. Thus the *maquis* can stand as a symbol for the continuity of the use of fire as a tool of environmental management (and accidental change) from the late Palaeolithic to the present.

It is impossible to consider the pre-industrial ecosystems of the region without thinking of the towns and cities. In AD 1500, the four largest European cities, with populations of 100 000 to 200 000, included three in Mediterranean lands (Naples, Venice and Milan), two of which are actually ports on that sea. In 1600, Naples had grown out of that size bracket, but additions to it included Rome, Palermo, Messina and Seville. Environmentally, the cities, as always, transformed land to a condition of higher energy throughput, and acted as concentrators of materials. Thus they often had industries which necessitated the gathering of fuel from a large area or the diversion of water-courses to power mills. In turn some of these manufactures (tanning, for example) produced wastes that contaminated rivers and estuaries. The urban population produced a demand for food, most of which had to be met from local sources, although salted and smoked meat and fish might be traded long distances. Local land use was therefore intensified by such demands, though there might be a contrary trend in the retention of pasture for the many horses; the richer the city, the more horses there were. The rich, too, inevitably wanted their country houses and estates outside but not too far from the town, as in the villas to the north of Florence or inland from Venice. Competition for the land immediately outside the wall increased when new ways of fortification required a bare area or *glacis* to give the gunners a clear fire zone and an area bereft of cover to discourage attackers.

In spatial terms, though, the impact of agriculture and pastoralism dominated most of the pre-industrial landscape. Agriculture requires land transformation: the layout of fields is a simple example which may reflect as many differing social and political pressures as it does environmental influences. The Roman practice of centuriation, i.e. of laying out fields chess-board fashion in squares of 20×20 *actus* (710 metres square), still underlies the active landscape in parts of the north Italian plain, and is detectable in air photographs of the now near-desert of the Tunisian Sahel.

That such a region should now be a near-desert raises interesting questions and ones to which there is no totally accepted answer. To what extent, it might be asked, is climatic change responsible for the greater aridity of the landscape? Or is the driving force that of human activity in using the land so heavily first for agriculture and then pastoralism that only bare soil subject to wind-blow and gulley erosion from the winter rains is left? Many studies have tried to tease out these strands for the late Roman and later periods, but few can reach conclusions which are sustainable over much time and space. It may have to be accepted that the interweaving of causal factors is so subtle and pervasive that we cannot separate out any single overriding cause. There is a rather similar situation in studies of the causes of desertification in the world today.

Irrigation too has a long history. It was certainly practised in ancient Egypt and Mesopotamia, important in Roman times and much extended in the sixteenth century as the towns grew. As a result, rice was added to the crop pattern and in effect a whole season (i.e. the summer) could be added to the cropping year. In the lands of the Mediterranean, however, irrigation is necessarily of limited use since there are few areas of lowland and plain. To cultivate the ubiquitous slopes requires another device, that of the terrace. Walls of mostly dry-stone construction are found on many slopes and provide a flexible resource: irrigated or rain-fed crops and fodder can be grown, and tiny pockets of production created out of impossibly rocky slopes. The opposite condition applies as well: there are places with too much water, such as in coastal marshes and lagoons, in river valleys liable to flood and in the great stretches of plain like the Po valley. All through the history of the region, attempts at drainage of such places were made; the early success of the Egyptians with clearing and irrigating the Nile lowland is well known. The Romans were highly active: the Pontine marshes south of Rome were tackled in 160 BC, and they started the long-term project of draining the Po valley, starting near Padua and Modena, thus producing land that the Italians called *bonifica*. As well as increasing crop land, such activities diminished the habitat for mosquitoes: malaria had been present at least in the eastern Mediterranean since 400 BC.

Irrigation in history provides us with a reminder that not all human-wrought changes are permanent. The example of the civilisations of Mesopotamia, watered by the Tigris and Euphrates, is instructive: unlike those of Egypt, they have proved impermanent since most of the land is now steppe and desert. What was present in their heyday (as when visited by Herodotus in the fourth century BC) was a complex set of dykes, dams and ditches, straightened rivers and storage lakes, with crops of wheat, barley and millet, sesame and groves of date palms. Such a system has to be carefully regulated. The earthworks must be constantly repaired and constructions such as sluices need maintenance, the quantity of water taken off by any individual must be subject to calculated and enforceable rules, and there must be a framework for coping with unusual circumstances such as droughts and periods of excess water. So, tight control by some central authority seems called for, whether this be mutual coercion mutually agreed upon (the socialist model) or imposed by an élite (the authoritarian model).

Why did these systems break down in Mesopotamia? There are various forms of explanation given:

1   The first set involves ecological change external to the irrigation system and might have had two main agents:

(a) A natural cause like climatic change, which if getting wetter could cause more erosion of non-forested watersheds, adding silt to the run-off of rainwater into rivers and thus clogging the irrigation system. If getting drier, then the quantity of water in the system might fall below that needed to support urban and other non-productive human populations.

(b) A human-induced change such as increased pastoralism on the watersheds of the major rivers. Heavy grazing shifts steppe and grassland vegetation towards a more desert-like condition in which soil loss is exacerbated and the resultant silt in the run-off will clog the irrigation channels.

2   A second set of possibilities might focus on autochthonous (locally induced) ecological change. The balance of evaporation and precipitation in this region lies heavily in favour of the former. Thus there are two effects in the soils which are the basis of the irrigated agriculture:

(a) waterlogging – there is an upward capillary movement of water in the soils, which is made worse if the canals are not kept clean because the water-table is then that much higher;

(b) salinisation – the water takes with it salts in solution which are then precipitated in the upper horizons of the soil profile, making it so saline that it may bear only a restricted range of crops tolerant of the higher salt levels.

3   The third explanation revolves round social and political process and relegates the ecological phenomena to effects rather than causes. If the control system breaks down then the irrigation system will no longer function since the discipline necessary for repairs, water usage, and channel clearing may no longer exist. War, whether civil or external, can only make such situations worse. In this version, therefore, the breakdown of the political system precedes the collapse of the ecology of the production system. A population used to the heavy hand of authority is unlikely to regulate itself even in the absence of armed conflict.

Given that the kinds of firm evidence about Mesopotamia are restricted to, for example, knowledge that there were periods of both political stability and instability, that there were times when barley crops increased at the expense of wheat (and barley is more tolerant of salt in soils than wheat), and that 'fossil' channels excavated by archaeologists show signs of silting up, what can we conclude? Only that it is dangerous to adopt simple cause-and-effect mechanisms as the explanations for events and processes long ago. Only if the evidence is secure, preferably of more than one type, and is clearly dateable into a chronological sequence, can we start to make the 'grand cause' statements so beloved of scholars of so many disciplines.

The fall of the Roman Empire is another popular topic for the seeking of grand-theory explanations:

1   Here again, we can start with ecological possibilities. Increasing levels of population and demand for food led to deforestation of watersheds in the mountains of the Italian peninsula. In turn, this caused flooding in the lower reaches of the valleys. Although some silt deposition might have enhanced fertility, it seems more likely that crop and stock losses undermined nutritional standards. Scarcity of timber also is a problem for urban and military-based empires.

2   There is then the misuse of technology possibility. The use of lead piping, especially into the homes of the rich, creates the likelihood of long-term toxification of water supplies. Lead poisons the central nervous system of humans, and children are especially vulnerable. So those nasty imperial habits were a result of lack of scientific and technological knowledge.

3   To which can be added the long-standing political and social changes of
the time. The pressures on the Roman Empire's solidity were growing, both
from nationalism inside and from enemies outside. The socio-political
fabric of Rome itself was not proof against all pressures and some of the
individuals with power were not perhaps the right people in the right place
at the right time: usually they are accused of moral decay and decadence.

But how is it possible to state firmly which of these processes is the 'cause'
of the fall of Rome? It seems unlikely that any one of them alone would
undermine the whole enterprise and we are left with the conventionally
unsatisfying even if more 'truthful' result that there was an inextricable
web of cause and effect which from the viewpoint of today cannot be
further teased out.

As elsewhere in the world, agriculture, pastoralism and the conversion
of forests to open land produced soil erosion. Many Classical writers
referred to this in rather stark terms, though it is by no means agreed that
their words referred to all of Greece, nor that the kinds of loss of soil that
revealed the bare limestone bones of the landscape went on throughout
history. Reafforestation and careful forest management were almost
certainly practised by some of the Italian republics: Venice, for example,
carefully looked after the forests from which its fleets of galleys were built,
not only in the Veneto but in Dalmatia and probably Crete as well. In this
respect they were following in the footsteps of the Emperor Hadrian (AD 117
to 138) who had decreed the protection of forests in the Lebanon, Syria and
Palestine. Metal-smelting and urban life both required a sustainable supply
of charcoal, which also led to woodland management rather than clearance.
Given, though, that there has been a long period of agriculture, and of
pastoralism aided usually by fire, given the steep slopes surrounding much
of the Mediterranean (especially the north coast), and given the
concentration of rainfall into a few months, then loss of soil and rock into
the rivers (causing flooding in winter) and into the sea (allowing siltation of
harbours and the creation of lagoons) was inevitable.

These processes were described in detail by one of the great pioneers of
environmental impact studies, George Perkins Marsh (1801–82) in his book
Man and Nature, or Physical Geography as Modified by Human Action
(1864).This work derived from his experience as a farmer in New England
and as a US diplomat in the Mediterranean region. Much of the book is
devoted to forests, because Marsh was an early advocate of what we would
today call watershed protection, which is ensuring that the litter and root
layers of the forest are able to act as sponges for water as well as retaining
soils and weathered rock. When the forests are cleared, much of the soil
might be lost. Subsequent deposition of the silt fraction on flood plains
could be a useful nutrient supply, but deposition of larger fractions (e.g. of
cobble and small boulder size) could render land sterile as well as doing
immense damage to structures and communications. So G. P. Marsh is very
much a father figure in the kind of study represented by this chapter. He
and later followers had considerable influence in reminding their readers
that the human domination of the earth was not complete and might not be
achieved (if accomplished at all) without paying a high price. From time to
time this message gets overlain by the more dazzling achievements of
science and technology but at present it is being actively canvassed by the
Green movement.

To return to our main narrative: not all land transformation in the
Mediterranean was in the service of food-getting. Quarrying might be
localised where there was good stone but then would exert a strong

environmental impact; for instance, Mount Pentelius bore a gleaming white
scar where a particular marble had been taken out. For hundreds of years,
Brac stone from a small island off the Dalmatian coast was sought for
buildings in the northern Adriatic region. Mining for silver, lead, copper,
mercury and arsenic have all been carried out around the Mediterranean,
with Spain an especially rich source. Surface and ground waters would
then become clogged with debris from the mines and the run-off would
almost certainly become laden with toxic substances. In the search for gold
in Andalucia in Roman times, hushing (the temporary damming of a
stream and then the sudden release of the head of water over the ground to
expose the mineral vein) was used just as it was in the northern Pennines of
England in the nineteenth century for lead.

In history, as now, pleasure was an important element of
Mediterranean life. On the land we see, for example, the creation of
beautiful gardens. The Egyptians may not have been the first, but their
elaboration of rectangular enclosures containing water, shrubs, flowers and
a summer house, and yielding fruit and fish as well as aesthetic qualities, in
some way set the tone for many later gardens – for instance the formalised
complexity of the gardens of Renaissance Italy, or the obvious conquest of
the hydrological cycle in the fountain courts of the Alhambra at Granada.
Utility and pleasure might also be combined, as in Classical Athens where
the great philosophers often taught in a garden: the Academe was the name
of the garden Plato used. Learning transferred indoors as it moved
northwards: first to the partially enclosed cloister and then to the
windowless lecture-hall of today. The cloister reminds us that religion at
various times would have protected environments from change, especially
woodlands. The ancient Greeks, for instance, had sacred groves, and the
Mount Athos peninsula today maintains unmanaged woodland as a matrix
for its secluded monasteries.

The lesson of these highly selected examples from the Mediterranean is
this: even in 1800 there would have been few places that were in any sense
natural. A few sand dunes, some mountain screes, cliffs, and perhaps the
deeper parts of the sea itself that were little fished, would have been the
only places that had not experienced change due directly or indirectly to
human agency. A striking feature is the polarisation of responses to the
changes. On the one hand, the poet Lord Byron (1788–1824) put it a bit
over-dramatically when he said that 'man marks the earth with ruin'. On
the other hand, many visitors have regarded Mediterranean landscapes as
the ideal. Further, it was the place in which the painting of landscape was
first practised, thus bringing the past into close proximity with any present.
It is a landscape which bares its past rather than burying it under two
metres of brown-earth soils. So the geographer J. M. Houston (1964) could
eloquently write: 'Mediterranean landscapes have the enduring qualities of
an eternal present, like beaches on which the tides of successive civilisations
have heaped their assorted legacies'. The juxtaposition of lush gardens,
farmlands, eroded hillsides and monuments of the past is illustrated in
Plate 12.

Given the perceived value of such paintings, and the sense of the past
made visible in the landscape, it is not surprising that so many people see
this region's environment even today (and certainly between the coming of
the railways and 1939) as evidence for the existence of a Golden Age of
mankind. No matter, perhaps, that such a stage never existed: the regional
relations between humans and environment was such that one group of
outsiders at least were able to identify it with one of the most enduring
myths that humanity has ever produced.

4   Industrialisation

4.1   Introduction

The essence of the changes which came to full flower in the nineteenth century, and which have borne their fruits ever since, is access to the stores of concentrated energy in the Earth's crust. Coal, oil and natural gas are collectively known as the fossil fuels (produced in the earth by fossilisation) and differ from the power sources of hunter–gatherers and agriculturalists in two main ways: they are much more concentrated and they are non-renewable. Because they are all so rich in energy per unit volume, a relatively small investment of effort in getting at them will bring a manyfold yield. Their use has been to generate steam to power machines, to fuel machines directly, to generate electricity, to make possible the chemicals and plastics industries, and to underwrite an increase in the world's population from perhaps 600 million in 1700 to the 6000 million of the present.

From 1800, therefore, it has been possible for humanity to transform the natural world (and the previously existing humanised world as well) in ways not previously possible. The new sources of energy can be directed via a versatile and ever-inventive technology at manipulating the ecosystems of land and sea, and the effectiveness of this process is greatly increased by the increased knowledge conferred by science. From an ecological viewpoint, industrialisation means that this new energy has been added to the solar flow, resulting in a subsidy that increases the intensity of output from systems that were previously powered by energy from sunlight alone. Agricultural productivity per hectare or per person-hour can be greatly increased: hence the ability to feed most of the extra population since 1800. The technology now available means that we can get at almost any part of the globe and alter it: the top of Everest, the South Pole and the floors of the deep oceans all show evidence of the presence of men or machines. Dredging of the oceans' floors for manganese nodules, the hunting of whales in remote waters, the placing of restaurants at the tops of mountains, the conversion of Pacific islands into quarries or airbases: all these are now possible. Given, too, that our wastes are often in gaseous or aerosol form, we can affect the whole planet and its atmosphere. All this is due to our command of the fossil fuels.

4.2   Origins and spread of industry

A phenomenon as complex as modern industry is difficult to define, let alone assign a birthplace. For the moment we will take as our criterion the idea of an economic activity powered by fossil fuels plus electricity from whatever source. For our spatial starting point we will take England of the eighteenth and nineteenth century as being the best candidate for the world cradle of industrialism, and the changeover from charcoal to coal products (notably coke) in the smelting of iron-ore as a key process. A charcoal 'bloomery' might extract 15% of the iron from the ore but the blast furnace will recover 94%. This technological development released the gamut of both qualitative and quantitative transformations that led to a fully formed industrialism in England by 1840. (The remains of the very early stages of this process are preserved in the Ironbridge Gorge Museum at Telford in Shropshire.)

From Great Britain, the new ways spread quickly to form what are now the core nations of the 'developed' world. By 1870 there were distinct industrial regions in France and Belgium, Russia, Germany, USA and Japan, all with the characteristic of being based on coal as a fuel, even if it had to be imported or even if, as in the case of the USA, wood and water had been the founding fuels of industrialisation. Along with industrialism went rising population levels. By 1880 the population of Great Britain, for example, had trebled in less than a century, and this experience was repeated elsewhere. Unprecedented, too, was the way in which these extra people were concentrated into towns and in emerging conurbations like the West Midlands of England and the Ruhr District of Germany. Once set in motion, industrial growth was for a while rapid nearly everywhere: in the UK energy use per person rose from the equivalent of 1.7 tons of coal in 1850 to the equivalent of 4.0 tons in 1919.

Outside these areas, the older ways persisted. The opening of the Stockton and Darlington railway to passenger traffic in 1825 or the launching of Brunel's screw-driven *Great Britain* steamship in 1843 probably went unremarked in Mindanao or in Addis Ababa but in time the echo of the steam whistle and the outer ripples from the docks in Bristol reached even such far-flung places. The railway, the steamship and the telegraph were to be the means by which industrialism was to be fed (in terms of primary products) and spread (in terms of trade, helped by gun-boats where necessary), and its concomitant ideas transmitted all round the world.

4.3   *Global impacts by the early twentieth century*

To organise this material, I will employ the concept of **core and periphery** on two scales: the impact of an industrial plant upon its local and regional environment, and the developed world upon the rest of the globe's ecosystems.

Starting at the local level, we can imagine a turn-of-the-century industrial plant such as a large coal mine or perhaps a coke works or steel mill. Figure 2.8 summarises the local environmental impacts, which are dominated firstly by the change in land use, with a large area now being covered with buildings, heaps of materials, roadways and yard surfaces, and transport features like railways. When the plant was under construction there would have been a great deal of bare soil which would have shed considerable quantities of silt to the run-off, but when completed the plant would show all the hydrological features of urbanisation, and especially the ability to shed water very quickly after rainfall, since the new surfaces absorb little or no water. A heavily built-up zone is therefore likely to make the local watercourses more 'peaky' in their response to precipitation. The quality of the water is also likely to have been affected since some of it will have percolated through waste tips, or have been used for washing materials or for cooling purposes. Its load of both suspended and dissolved matter is therefore higher, and some of the substances now carried are likely to be toxic to life. Water which now carries domestic and industrial sewage and sullage is also different in quality. The action of bacteria on the greater organic component of the sewage uses up more oxygen (i.e. there is a greater biochemical oxygen demand). Reduction in dissolved oxygen reduces or eliminates fish and

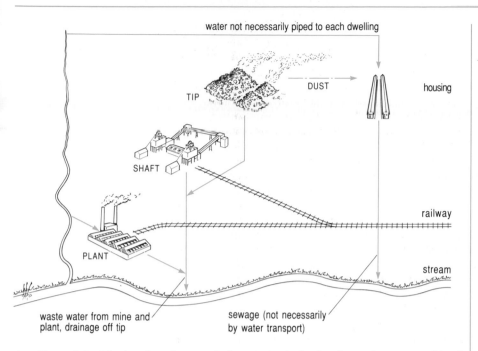

water not necessarily piped to each dwelling

DUST

TIP

housing

SHAFT

railway

PLANT

stream

waste water from mine and
plant, drainage off tip

sewage (not necessarily
by water transport)

▲   *Figure 2.8   The local environment of an early shaft mine for coal, an assembly of
structures that would probably have replaced agricultural land. The changes would
have included definite alterations of air and water quality.*

plants downstream from the outfall, so the watercourse is not likely to have
a great deal of life in it until the effluents have become well diluted.

Likewise, the air near the plant will be contaminated, and fall-out from
the effluent plume will depend upon distance from the source. Near the
chimney, particulates (soot particles etc.) will drop out and thus be a
disamenity and a health hazard to people living near their work: in the UK
the moneyed will try to live on the south-west fringe of a large industrial
area. Downwind, the effects get progressively less, but sulphur compounds
will rain out as dilute sulphuric acid and destroy building stones as well as
acidify the vegetation on wet uplands, for example.

This reaching-out of the effects of the plant is paralleled on the input
side by the tentacular stretch of the growing conurbations for water: local
upland valleys are likely to be submerged by reservoir construction. Each
plant is, however, linked more widely still through the raw materials it uses
and the sale of its products. The system of trade is what holds cities and
industrial plants together, and by the nineteenth century this was already
organised on a global scale.

If we extend this model to the whole planet, the developed areas of the
world can be regarded as the plant or the conurbation, and the rest as the
zones of outreach for materials or sinks for wastes. One of the chief
demands by the industrial zones was for food and so, for example, many
temperate grasslands were either ploughed up to provide cereals such as
wheat (e.g. the North American prairies) or converted to ranching to
provide meat, especially after the invention of refrigeration, as in Australia
and Argentina. Not all the foodstuffs were staples: tropical products such
as tea and coffee were in great demand and so large areas of forest and
bush were transformed into plantations. Colonial governments tended to
think of such lands as 'empty', though they may well have supported

seasonal populations of pastoralists or hunter–gatherers. As motor vehicles became more widespread, the demand for rubber increased greatly, so tropical forests in, for instance, Malaya were converted to rubber plantations, with sales of the original timber an additional source of profit. Many boreal forests began to have their ecology changed by the large-scale harvesting of trees for the pulp and construction markets during the later nineteenth century as well. Many other agricultural systems now found that growing for export was possible and so some changed their ecologies drastically by importing irrigation techniques, for example. We tend to think of the less developed world as major suppliers of mineral resources as well as crops, but in fact this was generally relatively small: in 1913, only six countries were major mineral suppliers to the industrial nations, compared with the 1960s when major industrial states took about 30% of their minerals from the tropics. But cheap transport by the steamship meant that foodstuffs could also be moved around the periphery: Burma (Myanmar), Siam (Thailand) and the Dutch East Indies (Indonesia) became major exporters of rice after 1870.

Another great impact of the industrialised regions upon the planet's resources was in the seas where steam trawlers facilitated the catching of fish as never before, and a map of the dates when particular fish populations became uneconomic to utilise is a map of (a) proximity to early-industrialised areas and (b) dates starting in the 1880s. Some populations recovered if left alone (e.g. in the North Sea during war-time) but in others the commercial species' place in the food webs of the oceans was taken by other species which might not be of interest to the trawlermen. The hunting of whales, too, was getting much more effective, with the aid of steam-powered mother ships and the use of explosives in heading harpoons.

▲ Whaling ship making a catch.

4.4   *The ecology of associated systems*

As we have seen above, the effects of industry go far beyond the factory gates. Although the city itself, for instance, is not solely a phenomenon of the nineteenth century and after, the growth of cities into conurbations is very much a feature of that phase. In environmental terms, large urban areas affect the environment rather in the manner described above for a single plant but on a much multiplied scale; what is more subtle is the way in which the ecology of agriculture in the developed countries has been transformed from a solar-based process to one which is underlain by large quantities of fossil fuel.

Earlier this century, this metamorphosis had not proceeded to today's extent (there was no sliced, wrapped bread!) but we can see the bringing on of the basic changes. In essence, this consists of the application in indirect form of fossil fuel to the food and fibre production sequence. Some of this takes place on the farm or other site of production: the use of powered machinery, for example. Steam engines were the first signs of this, in threshing, for example, and also in ploughing in North America. In both cases petrol and diesel tractors proved more flexible in use, though they replaced the horse quite slowly in Europe. (Wherever horses went out, a lot of land was freed from fodder production for other crops.) The tractor and other machines not only consume fuel on-site but represent a great deal of embedded energy used in their manufacture. This is true as well for chemical fertilisers, which began to replace manure in the pre-1918 period. The extension of cultivation into grasslands is facilitated by late-nineteenth century inventions such as barbed wire. Also, the winter of the temperate zones could be to some extent subverted by the early appearance of salad crops and spring flowers grown in metal-framed glasshouses. All this embedded energy is usually referred to as 'upstream' energy; 'downstream' from the farm, yet more energy is used in transport, storage (especially if the temperature is controlled to prevent spoilage) and

▲ *Modern combine harvesters.*

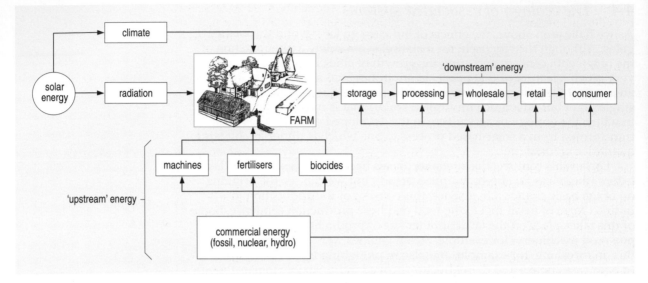

▲ *Figure 2.9   A simplified diagram of energy flows in modern agriculture. A number of processes supplement the energy of the sun in the production of food crops. The crops are then subject to a great deal of applied energy on their way to the consumers. In fact it is better to talk of a total 'food system' these days rather than 'agriculture'.*

processing (Figure 2.9). Thus wheat may be stored for some time at a controlled humidity achievable only by using a set of air pumps, and then it is milled using steam power and baked in an electric oven. This intensity of energy use was common early in this century but is low compared with that of today. The overall lesson is clear, however: the energy balance of the food system begins to move away from that of the surplus which must be produced in subsistence societies if they are to survive. Instead, a negative balance can be tolerated because fossil fuels can be mined to subsidise the whole process from agricultural college to table.

Q   The mention of agricultural college is a reminder that diagrams of energy or material flows, like Figures 2.9 and 2.5, omit a crucial aspect of farming systems. What is it?

A   Knowledge, and expertise in using it. Whereas pre-industrial farmers could rely on traditional practices to use the resources of the farm to produce significant output, the availability of new machines and chemical posed new management problems if farmers were to maximise profits. All modern societies put great stress on systems for educating and advising farmers. Unfortunately, prevention of harmful environmental impacts has become a priority only very recently.

The greater access to energy resources made available in the nineteenth century had its impact upon pleasure as well. In Great Britain, one environmental consequence could be seen (and still can) on the moorlands of the drier uplands of northern England and Scotland. Before the 1840s, it had been the practice of the owners of shooting rights to kill red grouse by walking them up with dogs and firing over the dogs at the rapidly retiring birds. But this changed; there was first the advent of the breech-loading shot-gun which could be reloaded very quickly and so allowed more targets per hour than the muzzle-loading equivalent. There was new money as well: the railways meant that the financier could leave his desk in the City of London at 4 p.m. and be on the Aberdeenshire moors next day at 10 a.m. willing to pay for a weekend's sport. But to go to these lengths to kill a couple of birds was clearly not economic. So a new system evolved: firing the moors to enhance the density of grouse. Since the breeding pairs are territorial, eat mostly tender shoots of heather and also require bushier heather for nesting sites, then it made sense to burn the vegetation to advantage the fire-tolerant ling *Calluna* (see also Chapter 6, Section 4.2). Virtually a 100% monoculture could thus be provided, in patches of different heights and bushiness. The grouse could then, come 12 August, be driven across the guns to be shot in huge numbers. All potential and actual predators were killed by keepers and people kept away during the breeding and shooting seasons. The firing of the moor at perhaps 15-year intervals meant that a great deal of its scarce nitrogen budget was lost in smoke, and soil erosion is always faster on such moors. Thus the long-term ecological process could be deleterious, just as the economics were advantageous.

The railways and steamships also began an era of mass travel for pleasure, of which today's inheritor is the package holiday to Spain or the Seychelles. Until the steam railway, a day at the seaside was something of an adventure for most, but great resorts sprang up in Europe, North America and Japan to cater for the day-trippers' desire for immersion in the newly healthy briny. The steam packet made the British marginally less insular since cross-Channel travel was not now so subject to wind ('Fog in Channel: Continent cut off ' ran, it is said, a *Daily Telegraph* headline in *c.*1900), and so the middle classes began to move out. One target for this migration was the Alps, very different from Britain in both summer and winter. Then, under the aegis of operators like Thomas Cook, really exotic places such as Egypt might be safely ventured upon: though the country was not as clean as Switzerland, the pound bought a great deal more there even in 1895.

So the ecology and energy balances of both necessity and pleasure had begun to change markedly by the time of the first world war, though their full course to our present situation had yet to be imagined, let alone experienced.

4.5   *Industry and environment in the Mediterranean*

Our choice of the Mediterranean as a source of examples is not perhaps the best for the earlier development of an industrial economy, though the region was vital in the development of trade and finance which preceded and underpinned industrialisation. As late as 1930, when the UK economy had only 6.0% of its working population in primary industries (production of materials by farming, fishing, mining and quarrying), Spain and Italy retained 48%, mostly in agriculture, though with some mineral extraction and fishing employment. As a result, their manufacturing sectors were small in comparison with the UK, Germany or the USA.

The persistence of colonial regimes around the Mediterranean, including the highly conservative Ottoman Empire, retarded the development of modern industry in all except the north-western shore nations of Spain, Greece and Italy, with the latter leading the transition to the more modern era. It seems that culture was the predominant element in the development of a new way of life because there are no factors of climate or natural resources which might have caused the delayed take-up of industrialisation. These places have after all now entered into such a phase, so the reason was mainly an unreceptive culture. In this case the dominance of colonial and imperial regimes was perhaps important in making sure that the Mediterranean lands provided labour and materials but did not get contaminated with new ideas like freedom and democracy which often seemed to go with factories and growing towns. Why do so many towns and cities from Paris southwards have wide boulevards in those parts developed or redeveloped in the nineteenth century? Civic pride maybe, but it is also very difficult then for insurrectionists to throw up barricades.

In the period being considered, Greece remained the most traditional in its economy, with some 10% more people in its primary sector than even Italy and Spain. In fact, it could be said that until the first world war, Greece experienced very little of the industrial revolution, one sign being a rate of population growth which showed little regional variation. There were a few exceptions, like the town of Volos in Thessaly, which grew from a population of 5000 in 1880 to 25 000 in 1907. The main factor there was the opening of three factories, two of which made agricultural machinery and the third processed tobacco, so that the link with the agricultural phase remained strong.

Of Spain, one economic historian has written of the 'failure' of the industrial revolution between 1830 and 1914. Through this period, the major evidence of it was probably the extensive working of minerals, mostly for export: lead, iron, zinc, mercury and coal were all produced in some quantity from the south-east Sierra, Murcia, Almeria, Vizcaya, Asturias and Cordoba. At one time it looked as though Andalucia might become a fully industrialised province, but this prospect had vanished by 1880 and the region became effectively de-industrialised. The exceptional area to this story is that of Catalonia, especially around Barcelona. Here, a local cotton industry provided the capital to develop chemical and metal industries, with the production in Barcelona of a static steam engine in 1849, a railway locomotive in 1854 and iron ships by 1857. Further industrialisation of this region, with the usual environmental changes from this change of economy, came with the construction of superphosphate plants in the late nineteenth century.

For a fuller penetration of industrialism into economy and environment we need to turn to Italy, and especially to the north. Just as cotton in Spain provided a link between a solar-based economy and a later coal-fired one, so in Italy the silk industry provides an analogous connection. From the sixteenth century, the demand for silk had led to the cultivation of the mulberry tree (to feed the silkworms) throughout Mediterranean France and Italy, and by the mid-nineteenth century there were 600 silk-throwing mills in northern Italy, mostly using water power. The industry then turned to silk weaving as well and to other forms of power and in so doing became the nucleus for other forms of factory production. There was, however, no iron and steel industry until the 1880s, when Genoa and Terni (Umbria) began production, and stimulated further demand for coal and iron-ore in places like Fiume and the Val d'Aosta, with consequent changes to air conditions and the rivers. Machine industries began to develop, mostly centred in established cities like Genoa, Naples and Milan, and then the transport industry opened in Milan and Padua with bicycles, cars in Turin (Fiat) and Milan (Bugatti) and rubber also in Milan (Pirelli in 1872). The auto industry in Italy at this time, however, was definitely a small-scale luxury-market affair and remained so longer than Ford, for instance.

The fast years of growth were 1897 to 1913 and one environmental consequence arose from the lack of cheap domestic coal. This was the development of the Alpine rivers for hydro-power. Italy was the first nation ever to transmit hydro-electric power (HEP) into a city, with a 5 kV line between Tivoli and Rome in 1892. Before that, HEP had been declared a public property in 1865, which led immediately to plants on the Adda and Ticino rivers, with larger plants coming in the last decade of the century; all of which can perhaps be symbolised by the fact that La Scala opera house in Milan was the first of its kind to be lit electrically.

So, by 1914 it would be possible to talk of a number of industrial concentrations which not only affected their local environment but caused change in their outreach for energy and materials as well. These were Barcelona, Marseilles, Genoa, Milan, Turin, Verona–Padua, and Leghorn–Pisa–Florence. In Italy, though, textiles were still very important in most of these, emphasising that the links with the earlier period were still in place. Italy was still somewhat sheltered from the brisk changes to the north: in 1870 there had been only two rail routes through the Alps (Figure 2.10) to her territory, but in the next 40 years another 10 were added. The main intention was to link Italy into wider European trade but one minor result was that Italy became opened up to the more northern middle-class culture-seekers, among whom the English have always featured prominently, as described in 1908 by E. M. Forster in the novel *A Room with a View*.

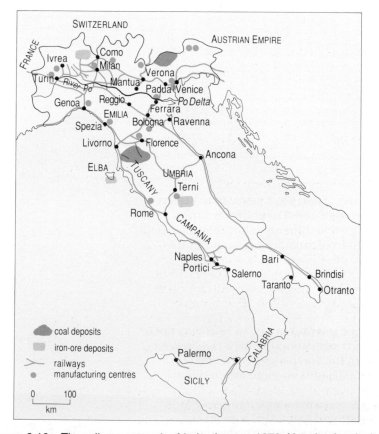

▲   *Figure 2.10   The railway network of Italy about AD 1870. Note its density in the far north and its sparsity south of Rome, with Sicily having only a couple of isolated lines. The extension to Brindisi provided a link with London via P&O steamers.*

Q   Bearing in mind what has been said in Section 4 about industrialisation and about the Mediterranean, why do you suppose that the industrial revolution first 'took off ' in Britain rather than elsewhere?

A   Several possible answers can be put forward:

●   Accounts used to stress the impact of mechanical inventions, but some technologies had been available elsewhere and not used, while others, for example in textiles, were more advanced in Flanders than in England.

●   Supplies of coal were vital to power pumps, machinery, trains and ships, but other areas like the Ruhr Valley had ample supplies of coal.

●   The conservatism of the Ottoman Empire has been mentioned as preventing industrialisation in the eastern Mediterranean. Some scholars have argued that Protestantism was more consistent with the growth of capitalism and industry than was Catholicism or Islam, and hence that northern Europe was advantaged over the south.

●   While all these factors may have played a part, it is also crucial that by the eighteenth century England had become the centre of the world economy through its dominance of trade. It thus had the capital and means of acquiring additional raw material as well as disseminating new products.

5   Conclusions

5.1   Introduction

Two kinds of conclusion need to be drawn. First, a summary of the degree and nature of environmental change brought about over the centuries. Second, a realisation that the facts are open to different interpretations, both in relation to the causes of particular environmental changes and in the value judgements that can be drawn.

5.2   The making-over of the world

Sufficient has been said to establish that the world in say AD 1800 was not in a pristine state of nature, and now it certainly is not (Figure 2.11). Neither was it totally a uniform human creation, for the changes were often regionalised, and some zones of the globe had been more affected than others.

The zones with the greatest changes were the temperate grasslands and forests, the lower slopes of temperate and tropical mountains, subtropical valleys and lower slopes and small islands. The temperate zone must be singled out as the most transformed because it experienced the immense changes caused by great agrarian colonisation movements,

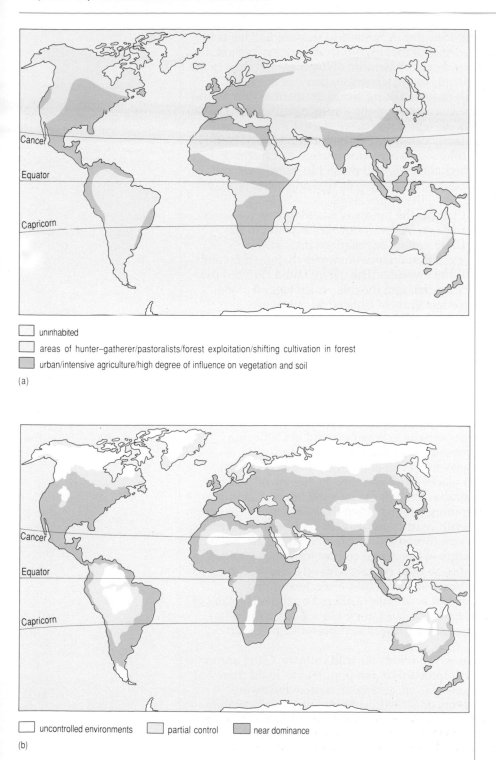

uninhabited

areas of hunter–gatherer/pastoralists/forest exploitation/shifting cultivation in forest

urban/intensive agriculture/high degree of influence on vegetation and soil

(a)

uncontrolled environments   partial control   near dominance

(b)

▲ *Figure 2.11   Estimates of the degree of environmental alteration of the land surface (a) by AD 1800 and (b) today. The 'uninhabited' category of (a) is replaced by 'uncontrolled' since anywhere (e.g. the South Pole) can be inhabited in a sense. 'Near dominance' in (b) means that human-organised systems predominate, though Nature sometimes comes back, as in earthquake zones, for example.*

though at different periods. Europe, for instance, lost most of its forests in the medieval period, whereas those of eastern North America persisted until the nineteenth century, and even now more of them are left than in the Old World. The natural grasslands of the temperate zone, as discussed above, were altered to feed the industrialising world. In warmer climes, the equivalent movement was the conversion of the forests of valleys and lower slopes to make rice *padi*, a process which may have started in the second millennium BC, but which certainly continued throughout recorded history. Temperate mountains, too, underwent vegetation change. For instance, in the Alps and Pyrenees the woodlands were converted to grasslands and were carefully managed in a communal fashion so that an animal crop was gained through seasonal grazing of the upper slopes.

Another site of rapid change was the far-away island where voyagers were apt to release pigs and goats on outward voyages in order to have fresh meat on the way back. These animal populations increased dramatically and the native vegetation underwent rapid change in favour of those species unpalatable to the invaders. The native fauna suffered too, not only from habitat change but also from the rats which tagged along with the domesticates; the flightless birds of isolated islands were especially affected.

Zones which were changed but to a lesser extent include the sclerophyll (i.e. hard-leaf, drought-resistant) vegetation biome, which encompasses the Mediterranean, the southern areas of the boreal forests and some of the tropical forest zones, especially the monsoonal forests. The sclerophyll zone of the Mediterranean saw its valleys converted to agriculture and the upper slopes either terraced for crops or used as grazing land; elsewhere in the world the pressures for change in these environments were less except perhaps in Australia and in California where, for example, fire suppression was practised and woodland succeeded chaparral (a low-growing scrub of evergreen woody plants). This part of California was in places much changed by the introduction of the eucalyptus tree from Australia. Though the lowland equatorial forests were less changed, their drier margins were often the site of much timber production (e.g. in the nineteenth century for teak from places like Burma and Thailand) and were also readily converted into plantations for palm oil and rubber. The southern margin of the boreal conifer forests, too, began to be more domesticated as they were used more intensively and replanted to fast-growing timber species. Many areas of coast were deliberately manipulated to form harbours and accidentally altered by the reworking of silt derived from land cover changes inland; their biota (flora and fauna) were often depleted by being choked with silt and poisoned by other wastes.

Lastly, there are the zones which were still wild country. Chief among these were the equatorial moist forests. Human populations lived there but they exerted low levels of manipulation, and external demands were mostly for selected species of tree or for wild rubber. Their role in global biogeochemical cycles, while not as it had ever been, was still more or less intact. At the other extreme of moisture, the deserts were still inhabited only by their nomadic and transient populations, together with a few Europeans captivated by the romance of the desert such as Charles Doughty, Freya Stark and T. E. Lawrence. Here, the contrast between oasis and arid was still firm, and the margins of the desert moved in response to climate rather than the multiple socio-economic interactions which seem to be at the heart of today's desertification processes. The higher mountains of the world were also left to nature. Climbing them had of course become a challenge to westerners whose world view had by then accustomed them to leaving nothing alone. But climbing Everest in tweed knickerbockers left no litter of discarded oxygen tanks *en route*; similar comments could be made about the various polar expeditions, which were rather less technological than most of their successors of recent times. The tundra, too, held little of interest to the west beyond an indigenous people who clearly needed missionaries, alcohol, sugar in packets and wheat flour, though not necessary in that order. However, the combined effect of sugared tea, bread, whisky and Sabbatarianism doubtless decreased the impact of hunting upon the animal populations, at any rate until the high-velocity rifle and the snowmobile came along. In every biome, many coasts were inaccessible from inland and still little changed.

It is difficult to paint a comprehensive picture of all these changes: perhaps they may be summed up by saying that by the end of the second decade of this century great changes had occurred and what was to follow was in many, though not all, places an intensification rather than an extension of such transformations.

Activity 4

Look at the list of impressions you made at the end of Section 1 of this chapter. Do they tally with this account? Did you, like most people, underestimate the extent of past transformations of natural environments?

5.3   Interpretations

Since the writings of Classical Greece, there have been people who are convinced that the present is a pretty bad deal and that the past was much better. Socially, for instance, there is nostalgia for the working-class terraced housing which was replaced by the tower blocks of the 1950s and 1960s; environmentally there is nostalgia for the Italy of the 1930s or the south of France at that period, when it was so different from industrial Britain.

For other interpreters, the perceived reality is different. The manipulation of ecosystems by societies can produce results which seem to possess positive features. They may be ecologically stable and hence sustainably productive. Some terraced land in the Mediterranean has been producing wine for thousands of years now with scarcely an interruption. Some small cities seem to have a magic combination of facilities, architecture and access to the countryside: Durham, for instance, or Siena. Other end-products of human use have aesthetic qualities which, despite the notoriously subjective nature of those judgements, command general acceptance as 'beautiful'. The port wine terraces of the upper Douro river, perhaps, or sunset over the Giza pyramids, or an autumn day in the remoter parts of the Cotswolds – these come to my mind but you can compile your own list. On the other hand, humans have produced processes to which the label 'negative' can be easily applied. The great irrigation schemes of ancient Mesopotamia became choked with silt, the soils became saline, and the great civilisations fell. It is difficult to discern cause and effect but the result was desert. In the course of industrialisation, the rivers of Britain became foul sewers. In the nineteenth century, sheets soaked in disinfectant had to be hung over the Thames-facing windows of the House of Commons so that Members of Parliament could continue their work. The beaches of many English and Spanish seaside resorts today bear visible signs of the lack of local sewage treatment.

The scale of these environmental processes, either positive or negative, is probably more important than we often realise; if large enough then whole societies can on the one hand be supported and on the other be threatened. For instance, scholars from time to time implicate environmental matters in large-scale declines of empires and civilisations, as with the Mesopotamian example given above. Generally speaking, the evidence from the past is never unequivocal enough to use the word 'cause' where environmental matters are concerned but a role is often found, as we saw in the discussion of the reasons for the fall of the Roman Empire.

Just as there are debates about positive and negative impacts of society on environment, so there are debates about the role of technology in producing those impacts. This chapter assumed as a starting point that access to energy was an important aspect of society's ability to transform environments, and subsequent sections have shown technology playing an increasing role. Interpretations of those transformations differ: some would see dramatic success for technology in providing for much larger human populations, many at unheard-of levels of material affluence; others would see the drama as a tragedy in which extraction of resources and increasing pollution has caused unprecedented damage to environments. You can make your own judgements about the balance of positive and negative impacts, but the explanatory point which needs to be made is that technology does not work independently. Technology has been advancing in societies in which many other political and economic developments were occurring. Industrial technology has been applied in societies where economic motives were increasingly central and perhaps only outweighed by military considerations. As the scale of trade, colonialism and ownership has grown, so those responsible for making crucial decisions have grown more distant from the environmental impacts of their actions, and perhaps less prone to take them seriously. But now those impacts have themselves become global: industrial society has transformed agriculture and even those few surviving hunter–gatherer groups, so that no environment is now free of human impact. The issue ceases to be one of causal interpretation and becomes one of finding ways to reduce harmful impacts. It means going beyond analysis to advocate action. Such action will undoubtedly call upon technology, but technology which is applied and controlled in new ways. The lesson of the past is that application of technology has been bound up with economic and political factors and influenced by cultural values. To change the impacts on environments may require a radical revision of the values of the society we live in, values which are explored in Chapter 3 of this book. Present global problems and systematic approaches to their interpretation are introduced in Chapter 4.

Environmental study can therefore be enjoyed at a number of levels. We can be engaged in a reform movement which seeks to alleviate environmental problems. At the same time, however, it is open to us to explore the fundamental nature of the world in which we live, both in its ecological dimension and in the ways in which we construct the world in our minds. This study is necessary if we are to continue as a species with the ability to realise its full potential, and in this exploration a knowledge of where we have been is an important element.

References

HOUSTON, J. M. (1964) *The Western Mediterranean World: an introduction to its regional landscapes*, Longman, London.

MARSH, G. P. (1864) *Man and Nature, or Physical Geography as Modified by Human Action*, Sampson Low, Son and Marston, London.

SIMMONS, I. G. (1993) *Environmental History: a concise introduction*, Blackwell, Oxford.

Further reading

DELANO SMITH, C. (1979) *Western Mediterranean Europe: a historical geography of Italy, Spain and southern France since the Neolithic*, Academic Press, London.

DI CASTRO, F. & MOONEY, H. A. (eds) (1973) *Mediterranean-type Ecosystems: origin and structure*, pp. 363–71, Springer, New York.

HUGHES, D. & THIRGOOD, J. V. (1982) Deforestation in Ancient Greece and Rome: a cause of collapse?, *The Ecologist*, **5**, pp. 197–207.

MANNION, A. M. (1991) *Global Environmental Change*, Longman, Harlow.

McNEILL, J. R. (1992) *The Mountains of the Mediterranean World*, Cambridge University Press, Cambridge.

SIMMONS, I. G. (1989) *Changing the Face of the Earth*, Blackwell, Oxford.

THIRGOOD, J. V. (1981) *Man and the Mediterranean forest: a history of resource depletion*, Academic Press, London.

Answer to Activity

Activity 3

Income	Expenditure
Cash	
£3820 (household food supply: beer, bread, bacon, lard, mutton, milk, butter and beef)	wages £740
	rent £900
	tithe £300
	poor rate £660
Kind	
107 ton wheat	food purchases (sugar, tea, salt and labourers' food)
142 ton barley	
690 sheep	
1400 fleeces	

Chapter 3   Changing attitudes to Nature

1   Introduction

In the previous chapter, Simmons outlined the growing severity and scale of human impacts on the environment as ways of life changed from hunting and gathering to agriculture and then to industry. Later parts of the course will show that those impacts have continued to grow as agriculture has been extended and intensified, as urbanisation, mineral extraction and energy use have accelerated and as even the oceans and atmosphere are significantly altered. Simmons also alluded to the beliefs and values which influenced the ways different kinds of society related to their environments. This chapter will focus on those beliefs and values in relation to **Nature** – the whole system of natural phenomena and plant and animal life. We have used the term 'attitudes to Nature' in the title to indicate that we are interested in dispositions or inclinations to behave in certain ways as well as in explicitly stated value positions. Indeed, much of the chapter seeks to make explicit the values of a range of societies, first saying a little more about hunter–gatherers and agriculturalists, then emphasising the values which underpinned industrialisation and the responses which it has provoked in the nineteenth and twentieth centuries. In conclusion, we speculate briefly about values for the globalised societies of the twenty-first century.

Evidence about the environmental values of past societies has been summarised by Oelschlaeger (1991). In brief, his argument is that hunter–gatherer societies do not separate themselves from Nature, but see themselves as a part of it. Typically, they had myths of origin that showed how they had originated from natural objects, and beliefs that regulated their use of resources. Many of these beliefs involved animism – the idea that springs, rocks, trees and animals were inhabited by spirits that had to be placated to prevent them unleashing punishments. With some exceptions, these rules contributed to the conservation of ecosystems over long periods, the maintenance of a usable harvest and the survival of this way of life for millennia.

In contrast with this idyllic picture, Oelschlaeger argues that the evidence is that agricultural societies did conceptually and practically separate themselves from Nature. Dependence on crops, and later on animals, led to a struggle against Nature, clearing natural ecosystems for fields and then trying to exclude wild animals, and resist weeds, pests and diseases. Religious responses to this situation soon included fertility cults and systems of gods who were responsible for major components of the environment – sun gods, river gods, forest gods and so on. As the societies grew more successful and sophisticated, these gods became separated from the environment and their interventions in the world more whimsical, as evident in both Greek and Hindu mythologies. Within the last few thousand years, agricultural societies developed the major religions which still survive, with some of the most successful recognising a single God located entirely outside the natural world. Some environmentalists

As you read this chapter, look out for answers to the following key questions.
- What have been the major historical perspectives on society and Nature?
- What new attitudes emerged in the 1970s and why then?
- What varieties of environmentalism are current in the 1990s and what are their strengths and weaknesses?

(following White, 1967) have argued that Judaeo-Christianity has predisposed European society to act in an irresponsible way towards the environment, especially through the Old Testament injunction to 'multiply and have dominion over the Earth'. Other scholars (e.g. Passmore, 1978) point to other strands of Christian thinking, one of which is discussed in detail in Section 2 below.

Plumwood (1993) traced another major influence on modern life. Greek philosophy, in setting out to replace superstition with a rational basis for right conduct, set up some conceptual categories which still exert an influence. As articulated by Plato, rationality was largely identified with masculinity and was contrasted unfavourably with irrationality which was associated with femininity, slaves and wild Nature. The highest ideals were located in the abstract sphere and the real world regarded as imperfect. Indeed, Plumwood saw the whole logical system as designed to glorify 'death in the service of the state' and hence to devalue life, women, aliens, Nature and feeling. Since Greek philosophy became, directly or via its influence on Christianity, a major influence on Western society for more than two millennia, these predispositions arguably have had serious consequences.

It is certainly the case that Greek ideas, as transmitted by Romans and Arabs, were an important contribution to the transition from feudal to modern industrial society in the seventeenth and eighteenth centuries. Social theorists now see this period as setting out a **modernist** agenda, which sought to throw off past religious and political restraints and to substitute a new era of progress based on rationality and humanism. This period saw the rise of science and democracy, but also of competitive individualism in pursuit of material gain. It was not long in historic terms before industrialisation became a dominant force in society, with the dramatic effects on ways of life and on environments which Simmons describes. These new problems stimulated a range of new philosophies and values, some of which are introduced in the next section.

After identifying the new philosophies and values stimulated by the industrial era, Section 2 concludes by relating them to the early conservation struggles described in Chapter 1. Sections 3 and 4 consider the forms of environmentalism current in the late twentieth century, but in the opposite logical order. Section 3 identifies some of the new concerns of practical environmental politics. Section 4 presents a classification which makes sense of these positions, and evaluates their strengths and weaknesses. Section 5 then argues that the new conditions at the turn of the millennium demand a perspective which is truly global, and assesses two current contenders.

Activity 1

Think about your own attitudes to Nature and environment: are they based on religion, science, literature, experience, or a combination of many sources? Are you aware of explicit rules governing human use of the environment or just of vague tendencies? Are you aware of contradictions within your own attitudes or between your attitudes and those which underlie 'progress'? How would you set about justifying whether one set of attitudes was preferable to a different set?

2   *The value of the natural world: some historical perspectives*

The idea we have of environment is a relatively new one, but it includes much of what used to be called nature and so questions about our relationship to our environment include questions about our behaviour to other life forms that share it. In this section we will be looking at four variously related sets of perspectives on the value of Nature. Since some of them are quite opposed, you will not be tempted by all of them. You should find, however, that through studying them you will be clearer about your own answers to the questions involved. You should also be able to recognise better the perspectives from which others approach these questions. You are quite likely to encounter something like each of these perspectives at some time or other. And it is as well to be aware that attitudes to the environment vary considerably. Deep-seated values and beliefs are involved on different sides of environmental debates. There is no point in arguing with others on bases they do not accept. There may even be limits to how far agreement can be reached about environmental issues. A study of a range of very different attitudes should give some sense of where such limits may lie.

2.1   *Stewardship of Nature*

The idea that the human race occupies a position of special privilege and responsibility in relation to the rest of Nature is one that appears to be endorsed by the first book of the Jewish and Christian Bible. The Book of Genesis incorporates this idea of a Stewardship of Nature in the story of Noah's Ark. There is a problem as to how this story should be interpreted but it is of interest from several points of view. It has been looked to, by Jews and Christians, as a source from which to derive a proper view of how humans should treat their environment. But it also of interest as evidence of an attitude to the value of the natural world that was held several thousand years ago.

The context of the story of Noah's Ark is a period of early Jewish culture in which, because of an ecological disaster or God's curse upon the land (or both!), Noah and his people had given up agriculture and turned to the life of nomadic herding and perhaps hunter–gathering. In such a culture it is to be expected that humans would want to preserve the species on which they depend. But an interesting point about the story of the Ark is that Noah takes himself to have a divinely ordained duty to save other species as well. Whether or not there ever was an individual called Noah who built an Ark (and so on), the author of the story comes over as a very early **conservationist** with a view about how humans ought to treat other species in the environment they share. The view is that these other species have a place in God's scheme of things and do not exist merely for humans to exploit. Humans, on the contrary, are accountable for their treatment of other creatures.

This view has been taken up within the Christian tradition in a number of ways. There have been, for example, those who took the diversity of species to be part of the richness or 'plenitude' that is a sign of the perfection of God's creation. Wantonly to destroy other creatures or cause the extinction of a whole species would be, according to this view, a religious offence. Such a view is embodied in one of the great poems by Samuel Taylor Coleridge called *The Rime of the Ancient Mariner*, first published in 1798 (Figure 3.1). Coleridge gave this summary of the 'argument' of the poem:

> How a Ship having passed the Line was driven by storms to the cold Country towards the South Pole; and how from thence she made her course to the tropical Latitude of the Great Pacific Ocean; how the Ancient Mariner cruelly and in contempt of the laws of hospitality killed a Sea-bird and how he was followed by many and strange Judgements: in what manner he came back to his own Country.

The Ancient Mariner, unlike the rest of the crew, survives the ordeal but suffers an 'agony' at intervals until he has forced someone to hear his tale. We are to imagine the poem as one such telling of the tale, to someone whom the Mariner prevents from attending a wedding. (The poem is included in its entirety in many anthologies.) *The Rime of the Ancient Mariner* follows the style of a ballad and deliberately incorporates elements of what its reader will readily imagine to be popular superstition. At one level the poem can be read as a superstitious tale about a sailor who is foolish enough to shoot an albatross – a bird of good omen – and the ill-luck that befalls his ship thereafter. But it can be taken at a deeper level as a warning to the reader of the dire consequences for humans if they fail to treat the

◄   *Figure 3.1 Illustration from a nineteenth-century edition of* The Rime of the Ancient Mariner *by Samuel Taylor Coleridge.*

lives of other living things with respect. In the concluding verses the
Ancient Mariner draws the moral of his tale:

> Farewell, farewell! but this I tell
> To thee, thou Wedding-Guest!
> He prayeth well, who loveth well
> Both man and bird and beast.

> He prayeth best, who loveth best
> All things both great and small;
> For the dear God who loveth us,
> He made and loveth all.

> The Mariner, whose eye is bright,
> Whose beard with age is hoar,
> Is gone: and now the Wedding-Guest
> Turned from the bridegroom's door.

> He went like one that hath been stunned,
> And is of sense forlorn:
> A sadder and a wiser man,
> He rose the morrow morn.

The Ancient Mariner story is in some respects different from that of Noah's
Ark. The Book of Genesis puts all living things under the 'dominion' of
humans and so puts humans in a position apart from the rest of Nature. *The
Rime of the Ancient Mariner*, on the other hand, presents the sailors and the
albatross on an equal footing as living things. All are wanderers across the
southern Atlantic where (according to Coleridge) living things do not
belong. The obligation of the sailors to the albatross partly arises from 'the
laws of hospitality', which the Mariner breaks when he shoots the bird.

In spite of these differences, both stories attach a value to the lives of
animals as such, both suggest human beings should treat animal lives with
respect, and both make humans accountable for their treatment of other
living things. Such Stewardship of Nature views are in marked contrast
with another view that, curiously, has also formed part of the Christian
tradition. Influential figures within that tradition (such as St Augustine, one
of the early Church Fathers) have tended to argue that human beings are
the only creatures that really matter. This view, sometimes aptly called
Imperialism, has been the dominant one within European culture in recent
centuries.

2.2   Imperialism over Nature

The attitude of Imperialism over Nature is clearly expressed in the writings
of Francis Bacon (1561–1626), who is regarded as one of the founders of
'Modern philosophy', a set of ideas and attitudes associated with the very
rapid growth of science in the seventeenth century. One of Bacon's most
important books was his *Novum Organum* (Latin for 'New Instrument'),
published in 1620. Bacon's purpose in this book was to advise his fellows
how mankind can 'regain their rights over nature, assigned to them by the
gift of God'. The reference to 'the gift of God' would almost certainly have
reminded Christian readers of God's instruction to Adam according to the
Bible: 'replenish the Earth, and *subdue* it: and have *dominion* over the fish of
the sea, and over the fowl of the air and over every living thing that moveth
upon the Earth' (Genesis I: 28, Authorised Version, italics added). This
might be taken as consistent with a Stewardship of Nature view. But,

rightly or wrongly, it was taken to license an Imperialist attitude to Nature in which everything else in the Creation is subordinated to the needs and wants of humans. The conquest of Nature, Bacon suggests, is the highest ambition a human being can have:

> It will, perhaps, be as well to distinguish three species and degrees of ambition. First, that of men who are anxious to enlarge their own power in their country, which is a vulgar and degenerate kind; next, that of men who strive to enlarge the power and empire of their country over mankind, which is more dignified but not less covetous; but if one were to endeavour to renew and enlarge the power and empire of mankind in general over the universe, such ambition is both more sound and more noble than the other two. Now the empire of man over things is founded on the arts and sciences alone, for nature is only to be commanded by obeying her.   (Bacon, 1620, Book I, para. 129)

The last words of this quotation have become one of the most famous Baconian maxims. To appreciate the force of this maxim – that 'nature is only to be commanded by obeying her' – it is necessary to bear in mind that 'obeying' Nature meant, in this context, reaching conclusions about the natural world only as a result of careful observation. This emphasis on observation rather than dependence on authoratitive sources or abstract argument, suggests that Bacon advocated a 'modern' view of science. This 'modern' view was for some time associated with the assumption that the aim of science is to control Nature.

The maxim that 'Nature is to be commanded only by obeying her' refers to Nature as female. It is not a peculiarity of the Imperialist view to think of Nature in this way. The idea of 'Mother Nature' is a very old one. But it is very likely that men's attitudes to Nature are affected by their attitudes to women. It has been argued (by Carolyn Merchant, 1980, Ch. 7) that this was so in Bacon's case, that Bacon drew an extended analogy between the interrogation of those troublesome women frequently accused of witchcraft in his time and the investigation of Nature. But that is an aspect of the matter that cannot be pursued here.

The Imperialist view of Nature was probably the dominant one in the eighteenth and nineteenth centuries and it still remains a common attitude towards the environment today. But there was already a reaction against this Imperialism on the part of some thinking people in the nineteenth century. A leading figure in the British reaction was the social critic Thomas Carlyle (1795–1881), whose prophetic article 'The signs of the times' was published in the then influential *Edinburgh Review* in 1829. Here is one passage in which Carlyle is, in effect, lamenting the dominance of the Imperialist view of Nature:

> . . . this age of ours . . . is the Age of Machinery, in every outward and inward sense of that word. We remove mountains, and make seas our smooth highway: nothing can resist us. We war with rude Nature; and, by our resistless engines, come off always victorious, and loaded with spoils.   (Carlyle, 1829)

Carlyle admitted that there were certain material benefits but believed that such an attitude towards Nature also has bad effects on those who hold it. Although he writes about how 'we war with rude Nature', he seems to think that we do not in the end make much impact on Nature itself. It is not the damage to Nature that seems to have worried Carlyle but the corruption of the human spirit that goes with attempting it. Carlyle expressed many of the attitudes towards Nature that were characteristic of what is called **Romanticism**, attitudes that have played an important part in shaping the conservation movement.

2.3   *Romanticism*

The word Romanticism is commonly used of a cluster of attitudes that were adopted by many artists, poets, musicians and writers from the late eighteenth century. These new attitudes were in many respects a reaction against the established attitudes of civilised Europe. As against the Imperialist view, according to which Nature had no value except what it was worth for humans, the Romantics accorded an intrinsic value to Nature. More originally, they attached a new value to wild Nature.

Wild landscapes had previously been regarded as dangerous, unpleasant and to be avoided. Keith Thomas, in his important book *Man and the Natural World*, offers this sample of earlier reactions to the mountains:

> Celia Fiennes [late seventeenth century] hated the Pennines and was glad to descend from the mountain rain to sunshine and the singing of birds. She thought the Lake District 'desert and barren' and its mountains 'very terrible'. In the 1670s Chief Justice North observed the 'hideous mountains' on his way from Carlisle to Appleby, while in 1697 Ralph Thoresby found both the Border country and the Lake District full of horrors: dreadful fells, hideous wastes, horrid waterfalls, terrible rocks and ghastly precipices.   (Thomas, 1983, p. 258)

The kind of landscape previously preferred by well-to-do English people is indicated by the following remarks, written in 1712, by the famous English essayist, Joseph Addison: 'British gardeners, . . . instead of humouring Nature, love to deviate from it as much as possible. Our trees rise in cones, globes and pyramids. We see the marks of scissors upon every plant and bush' (quoted in Passmore, 1978, p. 36) (Figure 3.2).

◄ *Figure 3.2*
Topiary at Levens Hall, Cumbria.

The famous landscape gardener of eighteenth-century England, Lancelot Brown, was nicknamed 'Capability' Brown because of his habit, when looking over the unimproved estates of his patrons, of describing the natural situation as 'having capabilities'. Edward Malins describes Brown as having sought 'to bring to life an improvement on rough Nature' (Malins, 1966, p. 66). This dissatisfaction with Nature in the raw seems to have been part of the common culture of the well-to-do in England in the eighteenth century. So, although they pioneered tourism in the Lake District, they felt the need of artificial aids to make the scenery pleasant for them (see Chapter 1, Section 3.3, 'Tourism').

The Romantic poets, such as William Wordsworth and Samuel Taylor Coleridge, did much to popularise a quite different attitude to the Lake District (see Plates 15 and 16 for pictorial representations of different attitudes to the Lake District). In 1803, after climbing the Kirkstone Pass in a storm, Coleridge wrote: 'The further I ascend from animated nature, from men and cattle and the common birds of the woods and fields the greater becomes in me the intensity of the feeling of life. "God is everywhere", I exclaimed' (Griggs, 1956–71, ii, 916). For many of the Romantics the cult of wild Nature was a kind of religion, called **pantheism**, in which God and Nature are identified. Wordsworth was also drawn to pantheism and expressed it, for instance, in 1798 in his well-known poem 'Lines composed a few miles above Tintern Abbey, on revisiting the banks of the Wye during a tour':

> For I have learned
> To look on nature, not as in the hour
> Of thoughtless youth; but hearing oftentimes
> The still, sad music of humanity,
> Nor harsh nor grating, though of ample power
> To chasten and subdue. And I have felt
> A presence that disturbs me with the joy
> Of elevated thoughts; a sense sublime
> Of something far more deeply interfused,
> Whose dwelling is the light of setting suns,
> And the round ocean and the living air,
> And the blue sky, and in the mind of man:
> A motion and a spirit, that impels
> All thinking things, all objects of all thought,
> And rolls through all things. Therefore am I still
> A lover of the meadows and the woods,
> And mountains; and of all that we behold
> From this green earth; of all the mighty world
> Of eye, and ear, – both what they half create,
> And what perceive; well pleased to recognise
> In nature and the language of the sense
> The anchor of my purest thoughts, the nurse,
> The guide, the guardian of my heart, and soul
> Of all my moral being . . .

Later in the same poem Wordsworth refers to himself as a 'worshipper of Nature'. Nature, for him, includes wild Nature but, as is appropriate for a Lakeland poet, he was also a lover of 'the meadows and the woods', which are usually human-made landscapes.

Wordsworth, as you saw in Chapter 1, played an important role in early attempts at conservation in the Lake District and ideas like his have provided, and continue to provide, an inspiration for conservationists. But many conservationists would draw the line at making the cult of Nature into some kind of religion. They may want to put their conservationist values on some kind of basis but not be happy about mysticism. They may prefer the relatively clear-headed and down-to-earth approach of those who are sometimes called hedonists and sometimes utilitarians.

2.4   The Utilitarian perspective

The basic idea of **Utilitarianism** or **Hedonism** is that the only thing that is to be valued for its own sake is happiness, pleasure or contentment. No beings matter in themselves for the Utilitarian except those that are capable of such states. Trees and flowers, assuming they have no feelings, do not matter except in so far as they affect beings that do have feelings.

The term Hedonism is misleading in that a hedonist is commonly taken to be someone committed to a rather narrow pursuit of the good life, interested only in those satisfactions that can be obtained from life without much effort. It was to avoid this slur that John Stuart Mill (1806–73) adopted the word' utilitarian'. This label is even more misleading, since the word 'utilitarian' is used of objects that are merely functional and give no pleasure. But the label 'Utilitarianism' is the one that has stuck. You are not likely to be misled by it if you bear in mind that it is an alternative to 'hedonism'.

Unlike any of the previous attitudes we have considered, that of the Utilitarian takes it as self-evidently wrong to inflict suffering on any creature capable of feeling it. But, if this seems obvious to us, it was not so when it was proposed by Jeremy Bentham (1748–1832) and others. Writing in a context where some progressive Europeans were beginning to wonder whether people should be allowed to inflict pain at will on slaves, Bentham foretold a time when people might no longer be allowed to inflict pain at will on animals:

> The French have already discovered that the blackness of the skin is no reason why a human being should be abandoned without redress to the caprice of a tormentor. It may come one day to be recognised, that the number of the legs . . . [etc.] are reasons equally insufficient for abandoning a sensitive being to the same fate.   (Quoted in Passmore, 1978, p. 114)

Bentham's principle was that the right action is the one that produces the greatest balance of pleasure over pain, that maximises pleasure or the absence of pain. Animals, in so far as they could experience pleasure or pain, came in for consideration almost on a footing with human beings. This was too much for some of Bentham's contemporaries and his Hedonism was labelled a 'pig-philosophy' for not discriminating between the higher pleasures of civilised human beings and those of animals. One of his followers, John Stuart Mill, tried to meet this criticism by giving greater weight to so-called 'higher' pleasures over the 'lower' ones. Mill's modified Utilitarianism, however, had much more of a bias to the human species, in which the place of animals is at best marginal.

Mill lived at a time when the enthusiasm for wild Nature (as we saw in the previous section) affected many people in England, especially amongst the better-off classes. He too seems to have shared this enthusiasm. But the value of wild places for a Utilitarian would, according to Mill, derive from the pleasure people take in being alone in them. This makes it a problematic pleasure for a Utilitarian to want to promote, since it cannot be maximised. Many pleasures, including higher ones (listening to music for example), can in principle simultaneously be enjoyed by any number of people. But one person can only enjoy being alone on the mountains if they are virtually alone. And that means other people cannot be enjoying the pleasure of being alone in the same place at the same time. The economist J. K. Galbraith has wittily remarked: 'A conservationist is a man who concerns himself with the beauties of nature in roughly inverse proportion to the number of people who can enjoy them' (quoted in Passmore, 1978, p. 105). This is a problem for utilitarian conservationists, who cannot on their own terms ignore the question of numbers. If the pleasure wild places give to people who like being alone in them were the only reason for conserving such places, it would be difficult to see how this could claim a high priority for a Utilitarian, compared with promoting pleasures that could be enjoyed by a greater number of people.

There are, of course, other kinds of reason Utilitarians might have for being keen on conservation, since there are many other kinds of benefit they will acknowledge than the pleasure of being alone in the mountains. We have mentioned this example because it illustrates how people (here Coleridge and Mill) with very different perspectives may none the less agree in particular cases. It might also serve to illustrate how people whose perspectives have certain similarities should not be too quickly lumped together since their disagreements in practice may turn out to be considerable. There are, for example, points of affinity between the Stewardship of Nature attitude discussed above and Romanticism, something that may be apparent from the fact that Coleridge was quoted as someone who was attracted by both. But the first attitude relates only to living things and not to rocks and cliffs. Coleridge, in enthusing about his ascent of Kirkstone Pass, thought (not entirely correctly, of course) he was getting away from 'animated nature'. So he was not thinking of the mountains of the Lake District as a habitat. And nor was Mill. Mountains could have a value for both a Romantic and a Utilitarian even if they were devoid of living things.

Activity 2

Compare the four perspectives on the natural world outlined above. Make notes of similarities and dissimilarities that may have struck you. Which perspective seems to you the most satisfactory? You will find it helps you to do this if you consider specific issues from the standpoint of each perspective as well as the stand you would be inclined to take yourself. For instance, why might it matter, if at all, if a species of flower in the Brazilian rainforest is lost as a result of the land being developed for more profitable agriculture? You will find comments on most Activities at the end of this chapter.

2.5   *Early environmentalism*

Look back at Section 3.3 of Chapter 1 and Box 3.1. What kind of environmental attitudes do you recognise as being at work in these cases?

In both the Lake District and Yosemite it is evident that the main threats came from environmental Imperialism in the form of capitalistic free enterprise. This was largely individualistic in the USA, with gold

> ### Box 3.1   *John Muir and wilderness preservation*
>
> Unfortunately man is in the woods and waste and pure destruction are making rapid headway. If the importance of forests is at all understood even from an economic standpoint, their preservation would call forth the watchful attention of the government.
>
> Thousands of tired, nerve shaken, over-civilised people are beginning to find out that going to the mountains is going home, that wilderness is a necessity and mountains, parks and reservations are useful not only as fountains of timber and irrigating rivers but as fountains of life.   (John Muir)
>
> John Muir was born in Scotland in 1838 and brought up on a farm in Wisconsin. In 1868, after an accident that nearly cost him his sight, he travelled to Yosemite in search of 'somewhere wild'. In fact the valley was already quite developed, with hotels, farms growing crops to feed visitors and horses, and a number of paved trails up the valley side. Muir used the valley as a base for long hikes into the high Sierras (mountains), where one of his early contributions was to recognise long before the academic geologists that the dramatic landscapes
>
> had been caused by glaciers. As well as solitary walking, he began to lead other walkers, became a celebrated raconteur and writer, and so communicated his experiences and sense of wonder to a wide public. He was vital to the campaign to have the 1200 square miles of high country designated as a national park in 1890. In 1892 he was asked to become president of a new club set up initially by academics from Berkeley – the Sierra Club. In 1903 he led President Roosevelt on a camping trip in Yosemite and persuaded him into a massive expansion of national parks, forests and monuments over the next few years. But though his successes were great, his career ended with a failure: in spite of vigorous campaigning, the preservationists were unable to prevent Congress granting San Francisco permission to dam Hetch Hetchy – the second major valley in Yosemite Park. The city's arguments were supported by Gifford Pinchot, the United States Chief Forester and advocate of conservation interpreted as 'wise use'. Even the Sierra Club was divided by the issue and in the atmosphere of sympathy for San Francisco after the 1906 earthquake, permission was given and the dam built. John Muir died, some say of a broken heart, in 1914. (Adapted from Sarre, P. *et al.*, 1991.

prospectors, homesteaders, ranchers and hotel owners in evidence, but more corporate in the UK as coal, iron and railway companies sought to exploit the resources of Cumbria.

In the case of Thirlmere and of Hetch Hetchy, the development interest had a more Utilitarian flavour. Here, it was municipal governments seeking additional water supplies to allow them to provide better services, and hence improved hygiene, to urban dwellers, including the poor. So these proposals could be seen as socially progressive, as aiming at 'the greatest good for the greatest number'. This put the defenders on difficult ground: although the term had not yet been invented, they would have been marginalised as NIMBYs – protecting their own interests by opposing development (not in my back yard!). Perhaps that is what forced them to seek for innovative forms of self-justification.

Wordsworth started from his own aesthetic and spiritual responses but sought to bolster them (surprisingly from an intellectual revolutionary and admirer of the French Revolution) with some rather conservative arguments. His positive feelings towards the countryside, traditional buildings and established communities are rather typical of British conservatives, though the emphasis on the aesthetic élite distinguishes his form of Romanticism from the more earthy pursuits of the gentry.

Muir was much more interested in contact with, and understanding of, wild Nature than was Wordsworth. At times he writes as if Hetch Hetchy, or even alligators, were worth preserving for their own sake – to the detriment of humans if necessary. He seems to have gone further towards ecocentrism (see Section 4) than other environmental activists of his time, possibly as a result of reading Thoreau and oriental philosophers as well as his own direct contact with the Sierras. However, he was politically astute enough to recognise the need for a political lobby for wilderness and wrote his books, entertained President Roosevelt and helped build up the Sierra Club in order to create that lobby. His success in building up support for preservationism shows that a wide public was interested in the spiritual and ecological benefits of wilderness even a century ago.

As the Sierra Club matured and worked increasingly in the corridors of power, it began to adopt a secular version of Stewardship. In this it was similar to the National Trust in England, which more directly established itself as the keeper of open land and country houses for the benefit of society at large and, as its insistence on inalienable ownerships demonstrates, for later generations. The spectacular expansion of the National Trust's membership and property portfolio in its first century demonstrates that its respectable, not to say aristocratic, use of inalienable property, gifts, legacies and public subscription is successful in preserving special cases.

None of the environmental attitudes discussed above does more than defend some remnant of wild or settled landscapes from development. In effect, they seek only to steer development, whether nakedly Imperialist or avowedly Utilitarian, away from favoured areas. They do not seek to challenge the nature of development itself. One nineteenth-century figure who began to do that was Ruskin (Chapter 1), who objected to the changes to the labour process under capitalist industrialisation. His own solution was backward looking, seeking a restoration of aesthetics and craft skills through a return to medieval guild organisation, but he helped inspire William Morris (1834–96), whose writings and political campaigning advocated a form of **socialism** (see Box 3.2 later).

Ruskin's and Morris' concerns about the nature of individual work reflected one of the concerns of the foremost critic of industrial capitalism, Karl Marx (1818–83). In some respects Marx was philosophically more aware of Nature than were conservative thinkers: he rejected any separation between society and Nature, arguing that human skills are natural and that wild Nature has been transformed over millennia into a 'second Nature'. However, his central concerns were social: he saw capitalism as extremely productive, but as alienating and exploiting labour, substituting market value for usefulness and out of human control. His advocacy of common ownership of the means of production was intended to restore control, to direct production to satisfy human needs and to provide more free time for workers to pursue their own happiness – a goal reminiscent of Utilitarianism. Unfortunately his theory saw the material of Nature as having no value until worked on by human labour. The idea that value could only be created through production encouraged the Soviet Union to embark on the transformation of society through industrialisation, collectivisation and ambitious programmes of water management. As a result, Soviet, and later East European, industrialisation had severe impacts on environment as well as failing to reach its social goals.

By the end of the nineteenth century socialism had become the main focus of criticism of industrial societies. In the twentieth century the only political movement of comparable importance was the pursuit of independence in the colonial world. Not until well after the second world war did the environment become a major issue again.

3   *The environmental movement*

Environmentalism, as we now understand it, is a relatively new
phenomenon, dating from the late 1960s and early 1970s. Like all cultural
phenomena, its roots lie in the past. Past attitudes to the environment
remain, as we shall see, at least partially available to new generations. But
the environmental movement was in some respects a departure from the
long-established conservation movement. This section considers various
kinds of evidence about the nature of this new environmentalism and some
aspects of its development.

3.1   *A new language*

To ask 'What is environmentalism?' is to ask a question about the meaning
of a word. As it happens, the word 'environment', in the sense in which it is
now familiar, is relatively new. The words 'environmentalism' and
'environmentalist', as they are now used, are even newer. Sometimes, of
course, changes in the meaning of words have no particular significance.
But sometimes they can be surface indications of significant changes in how
a number of people think. That is how it seems to be in the case of the word
'environmentalism'.

One way of tracing changes in the development of widely held ideas is
by looking at dictionaries prepared at different times. Dictionaries revised
after the early 1970s record, as we would expect, that an 'environmentalist'
is someone who is concerned about and works for the protection of their
environment. But until that time an 'environmentalist' was someone who
subscribed to one side of a long-running debate in academic psychology –
someone who stressed the importance of social circumstances rather than
hereditary factors in determining human personality. It seems to have been
in the late 1960s and early 1970s that the word 'environmentalist' acquired
its familiar meaning of an activist about environmental issues such as
pollution. The 1972 *Supplement* to the *Oxford English Dictionary* still gives
precedence to the use of the term in academic psychology – 'one who
believes in or promotes the principles of environmentalism', viz. 'who
considers that environment has primary influence on person's or group's
development' – but it adds to this meaning: 'also, one who is concerned
with the preservation of the environment (from pollution etc.)'.

Changes in linguistic practice are happening all the time and in most
cases they do not signify either the development of an important idea or a
change in social attitudes. There is, however, a good deal more evidence
that in this case they do, as we shall see shortly. But in any case
'environmentalist' was not the only word used to refer to those who were
caught up in these new attitudes. The word 'ecologist' had previously been
used almost exclusively of a specialist within the discipline of biology. But
it came to be extended, for a while, so as to refer more broadly to the same
sort of people as were being called 'environmentalists' in the new sense.
Thus E. M. Nicholson, in a book called *The Environmental Revolution*, wrote:

> . . . ecology is the study of plants and animals in relation to their
> environment and to one another. But it is also much more than that: it
> is the main intellectual discipline and tool which enables us to hope

that . . . man will cease to knock hell out of the environment on which
his own future depends. (Nicholson, 1970)

Quoting this passage, the author of the entry on 'Ecology' for *The Fontana
Dictionary of Modern Thought* (Bullock & Stallybran, 1977) refers to
Nicholson's use of the word ecology as 'debased'. But, in spite of such
purist disapproval, this broader use of the word did establish itself to some
extent in the late 1960s and early 1970s. One of the leading journals of the
new environmentalism, started in Britain in 1970, is still called *The Ecologist*.
There was also a British political party started in 1975 calling itself 'The
Ecology Party' (though it later became the Green Party). Although the word
'ecologist' is still used in this wider sense, it is the word 'environmentalist'
that came most into vogue. What is called 'Deep Ecology' refers to a form of
radical environmentalism in which fundamentally different human
lifestyles are envisaged. The phrase 'ecology movement' seems to refer to
such radical environmentalists.

   The word 'environmentalist' came to be used of quite a broad range of
people, including lay activists as well as professionals. One of the hallmarks
of the new environmentalism is the involvement in it of such lay people,
who have taken an interest in a wide range of environmental matters, often
of a quite technical sort. This lay readership came increasingly to be
provided for in the 1970s, for instance by the publication of environment
handbooks and dictionaries of various kinds. The publishers of one of
these, the *Macmillan Dictionary of the Environment* (Allaby, 1988), make it
clear in their blurb that they are responding to what they saw as 'the
popular environmental movement of the 1970s'. The fact that such
publications evidently sold well is further evidence of this lay interest and
of the growth of the environmental movement.

Activity 3

Look up dictionaries available to you, either at home or at your local
library and see if you find the following pattern confirmed: before the
early to mid 1970s it is the older meaning of 'environmentalist' that is
given precedence; after that only larger dictionaries even mention the
older use.

3.2   *The environmentalist organisations of the 1970s*

There are other kinds of evidence that a new environmental movement was
under way in the late 1960s and early 1970s. One is that new kinds of
environmental organisations were created about that time. Greenpeace and
Friends of the Earth are particularly worth considering, partly because of
their subsequent importance amongst the world's environmental
organisations and partly because their methods of campaigning were from
the start very different from those of traditional conservation bodies.

   Typical of conservation bodies established in the late nineteenth
century were those concerned with the protection of a natural heritage –
such as the Sierra Club in America, with its work to protect the wilderness
areas, and the National Trust in Britain, whose work in the Lake District
was referred to in Chapter 1. Other bodies had, at least initially, a more
specialised interest in conservation: for example the Audubon Society in the
United States and the Royal Society for the Protection of Birds in Britain.

The new environmental organisations were concerned with issues other than simply conservation.

Greenpeace was founded in Vancouver in 1970. Its very name indicates its association with the 'peace movements' of the 1960s and initially it was a small group who sought to prevent nuclear testing in the Pacific by sailing into the danger zone. It also became concerned with trying to prevent whaling by similar 'direct' methods of non-violent confrontation (Figure 3.3). It subsequently mounted campaigns to protect the North Sea, the Great Lakes and Antarctica. By 1995 Greenpeace had established itself as one of the best-known environmental organisations, with 4 million members and 40 national branches. One of the principles to which it has been committed is the equal right of all species to exist and flourish, a principle affirmed in its forthright advertising campaign of the mid-1980s against the wearing of fur coats.

Friends of the Earth began in the United States as a splinter group from the Sierra Club in 1970, and became involved in a large number of environmental issues in 28 different countries. A branch was established in London in 1971 and by the late 1980s there were around 280 local groups in

◄ *Figure 3.3*
Greenpeace boat
confronting a whaler.

Britain. Like Greenpeace, Friends of the Earth has close links with the anti-nuclear movement. It also sometimes adopts direct methods, first hitting the headlines when some of its activists dumped 1500 throwaway bottles on the doorstep of Schweppes (Figure 3.4). Jonathon Porritt, then its Director in Britain, wrote about the Friends of the Earth in 1987:

> Thousands of people have participated in consumer pressure campaigns, protests against acid rain, direct action to stop the destruction of irreplaceable wildlife sites, public meetings to stop nuclear waste dumps, cycle rallies, and many more. In addition, Friends of the Earth has published reports, promoted legislation in Parliament and participated in public inquiries.   (Porritt, 1987, p. 157)

We are not suggesting that these organisations constituted the new environmental movement of the early 1970s, only that they were one manifestation of it. In the United States, in particular, there was a good deal of public concern about pollution. On Earth Day in 1970, as many as 20 million Americans are believed to have taken part in demonstrations against pollution, particularly against the smog caused by motor car

◄ Figure 3.4
Throwaway bottles left
outside Schweppes by
Friends of the Earth.

exhausts in American cities. This public concern had some effect on the American government, which had already begun special activities to improve environmental quality in the late 1960s. The US Administration set up the Environmental Protection Agency in 1970 to promote environmental quality. Official attitudes seem also to have undergone a considerable change around this period. As well as the radical environmentalism of Greenpeace and Friends of the Earth there was what may be called an establishment environmentalism.

3.3   Establishment environmentalism

There are many sorts of evidence of the attitudes to environmental issues taken by those who saw themselves in positions of public responsibility. Official statements are obviously relevant but are apt to be too brief, bland or impersonal. Slightly less obvious sources may be more informative. Encyclopaedias, which enjoy a particular status in British and American culture as a means of authoritative communication, are one such source.

During the 1970s encyclopaedias gradually adopted the classification 'environment' for those topics they previously included under various headings, particularly 'conservation'. One of the first to make this change was the American *Merit Students Encyclopedia*, first published in 1967, but updated by year books. This encyclopaedia originally offered only a loose definition of what an 'environment' is ('the surroundings in which an organism lives') and the entry for the 1967 edition is very brief, identifying the term as belonging to ecology, itself treated as a branch of biology.

It would, of course, be remarkable had any such major reference work failed to include anything that would directly interest people who would nowadays call themselves 'environmentalists'. The original 1967 edition does include such material, under headings like 'conservation', 'erosion', 'forest', 'mining', 'soil conservation', 'water supply' and 'wildlife conservation'. These articles would probably seem to modern environmentalists to add up to an unsatisfactory, because limited and fragmented, treatment of their topic. And certainly the editors of the *Merit Students Encyclopedia* themselves seem to have acknowledged this. From 1971 the Year Books subsumed everything previously classed as 'conservation' under the head 'environment'. The rationale for this change seems to have been provided by the entry for the previous year under the heading 'Conservation and Natural Resources':

> As the decade of the 1970s began, the United States faced grave problems and marked significant achievements in the management of its natural resources. Above all, there were dramatic changes in the conservation movement. Even the word 'conservation' was falling into disuse in favour of more descriptive and more inclusive phrases. Educators and resource managers spoke of total environmental management and envisioned a discipline that would create a harmonious relationship between man and his environment. Under this new concept, the conservation philosophy was becoming less a tool of the physical sciences and more the domain of the social sciences. It stressed physical law less and human behavior more; it was less compartmentalized and more comprehensive, less local and more global, less rural and more urban and more people-centred.
>
> But the greatest change was in public attitudes. Anxiety about the US resource base was no longer the exclusive concern of the professional resource manager and the dedicated conservationist. An

opinion poll sponsored by the National Wildlife Federation showed that 51 per cent of all Americans were deeply concerned with their environment and that 35 per cent were somewhat concerned. Seventy-three per cent expressed a willingness to back up their concern with additional taxes. The reason was clear. Environmental mismanagement had reached a point where it could be recognized by the man in the street. The citizen who could accept with equanimity the extinction of the passenger pigeon or view the plight of the whooping crane without undue concern could not tolerate the thought of sewage in his drinking water or exhaust fumes in his breath of morning air.

Pollution was the greatest single environmental issue before the American public. There was general consensus [amongst environmental scientists] that environmental pollution, like other national conservation problems, was the direct result of two factors: the unbridled growth of human populations and the affluence of American society. In short, not only were there more people to make demands on the resource base, but each individual was demanding a higher standard of living, using more goods, and creating more waste.   (*Students Year Book*, 1970, pp. 195–6)

The author of the article from which the above extract has been taken, Paul Herndon, was at the time a writer–editor for the Information Office of the Bureau of Land Management of the US Department of the Interior, someone who was professionally entrusted to express officially authorised attitudes. The likelihood is that his statements are all based on impressions formed from reading, attending conferences, and so on, and that they do indeed reflect a significant change in thinking about environmental matters, particularly the kind of expert opinion with which the Bureau of Land Management would be in touch. The rather garbled summary of the change suggests that he is reporting, as he implies, what was envisaged by educators and resource managers who spoke of a new discipline of 'total environmental management'.

Activity 4

Consider how far the 'dramatic changes in the conservation movement' referred to in the above extract from the Students Year Book (1970) are exemplified by the emergence of organisations like Friends of the Earth. Make a note of the differences and similarities that seem to you significant.

3.4   The growing international sense of 'ecological crisis'

The Year Book article on conservation mentions some of the respects in which the environmentalism of the early 1970s was new. But it does not express the keen sense of human vulnerability to environmental changes that is characteristic of these new organisations. Even its editorial foreword, which highlights the emergence of a new concern about the environment as one of 'the . . . outstanding developments of 1969', still presents the problem as if it were a failure of stewardship, as if the Earth were something separate:

Burdened with an ever-expanding population, polluted by the products of man's ingenuity, and scarred by the heavy tracks of progress, the earth's future is in doubt. Increasingly, scientists have begun to fear for it, and their fear has been translated into constructive action by a community of concerned citizens.
(*Students Year Book*, 1970)

Pictures of the planet Earth as seen by the Apollo 8 astronauts during their orbital flight around the Moon in 1968 (Plate 3) contributed to a popular perception of the relative smallness and fragility of the Earth. However, the idea that humans are set apart from Nature – a feature of many (though not all) traditional attitudes towards the environment – seems to run so deep that even when people contemplate a global **ecological crisis**, as it is sometimes called, they sometimes resist the implication that the human race would be caught up in it too.

None the less, there was, in the early 1970s, a growing sense that environmental problems were not to be left to charities or individual governments but that they urgently required international co-operation. The year 1970 was declared to be European Conservation Year, though there was so much talk and so little practical outcome that critics dubbed it 'European *Conversation* Year'. The United Nations achieved greater success with its *Conference on the Human Environment* in Stockholm in 1972. This conference produced a declaration of principles, a plan to deal with certain specific problems and an agency called the Governing Council for Environmental Programs. The United Nations Environmental Program (UNEP) initiated a monitoring system to detect global pollution that might affect weather and climate.

3.5   Britain and 'the ecological crisis'

The new environmentalism, as a popular movement responding to a sense of 'ecological crisis', seems to have taken root firstly in North America. But it did not take long to spread to Europe. 'Environment' had become something of a vogue word in some circles in Britain by the early 1970s. A new government 'Department of the Environment' was set up in 1970. According to the White Paper on *The Reorganisation of Local Government*, this new department would be 'responsible for the whole range of functions which affect people's living environment'. It would be concerned, for instance, with the planning of land use and all matters to do with housing and transport. But, the White Paper goes on to state: 'There is a need to associate with these functions responsibility for other major environmental matters: the preservation of amenity, the protection of the coast and countryside, the preservation of historic towns and monuments, and the control of air, water and noise pollution.' (White Paper, 1970, Section III: 31). Clearly it was the intention to deal more effectively with what were then seen as environmental problems. But the new Department was not in any way the result of any official sense of ecological crisis. Even the word 'environment' is defined in the White Paper not in its ecological sense but very loosely as 'where people live, work, move and enjoy themselves'.

The new environmentalists, by contrast, tended to think of an environment in biological or ecological terms, and this involved a new sense of the vulnerability of human beings, like other organisms, to changes in their surroundings to which they may be unable to adapt. This sense of human vulnerability was expressed in a special issue of *The Ecologist* for 1972, entitled *Blueprint for Survival*. The authors of this manifesto believed that the rate of increase in pressure that humans were putting on their environment could not be sustained. They declared that 'if current trends are allowed to persist, the breakdown of society and the irreversible disruption of the life-support systems on this planet, possibly by the end of the century, certainly within the lifetimes of our children, are inevitable'. They outlined a strategy involving as little ecological disruption as human needs required and maximum saving of energy and other resources.

At a more popular level the sense of ecological crisis was sharpened both by concern over the threat of nuclear warfare but also by concern about atmospheric pollution being caused by nuclear tests.

CND and reactions to the supposed ecological crisis

If there was any group in Britain in 1970 that had the kind of sense of ecological crisis characteristic of the new environmentalism it was one that enjoyed little sympathy in Whitehall – the Campaign for Nuclear Disarmament (CND). I mentioned earlier that some of the organisations symptomatic of the new environmentalism (Greenpeace and Friends of the Earth specifically) had affinities with the peace movement. It has been suggested indeed that the 'grass roots' support for CND was fundamentally what we would now call 'environmentalist'. Writing in 1970, H. F. Wallis made the following claim about CND:

> It had a long list of aims, but what seemed to matter to most of its supporters were the risks to health, and especially to the unborn, by the release of strontium-90 into the atmosphere through nuclear explosions. Evidence that this was the major preoccupation is contained in the fact that after the test-ban agreement was signed, CND's supporters melted away and the movement has never been the same since. It may therefore be regarded as the first mass anti-pollution campaign we have seen in this country.
> (Wallis, 1972, p. 132)

Whether or not he was right about the reason for the drop in CND's support, it is clear that Wallis's sympathies for the CND were limited. He is strongly pro-conservation and anti-pollution in this book, which was intended to serve as a 'conservation handbook and directory' for campaigners to make Britain a better place to live in – which is why it is entitled *The New Battle of Britain*. But its author sees the problems neither as global nor as at a state of crisis. 'There is', he says in his concluding remarks, 'no sense in being unduly alarmist' (Wallis, 1972, p. 137). The phrase 'unduly alarmist' is directed against those who talked in what struck the author as an exaggerated way about an 'Eco-Crisis', as some had done during European Conservation Year, 1970 (Figure 3.5). It is probably also directed against the CND.

It seems that the new environmentalism was already in the air in Britain by 1970 and that H. F. Wallis expresses a reaction to it from the standpoint of a committed conservationist. Was his response representative of conservationists in Britain at that time? It is likely he believed his sensible, practical approach would appeal to many who wanted to work for environmental improvement. Talk of global environmental crises was still something of a novelty. It was not wholly new in 1970. In the early 1960s an American biologist, Rachel Carson, had argued, in her strikingly titled book *Silent Spring*, that the indiscriminate use of insecticides not only involves the inadvertent killing of animals and birds, but can even affect human health. In a preface to the English edition of this book, the biologist and

"ALARMIST!"
...THERE'S STILL 500FT TO GO

▲   Figure 3.5
Cartoon from The Ecologist, October 1972. There was some controversy at the time about how serious the ecological crisis was.

writer Julian Huxley prophesied that 'man is progressively ruining and destroying his own habitat' (Carson, 1965). We can see now that there is a real risk of this prophesy turning out to be true. But much of the evidence of such a global crisis that we would be likely to cite (such as the compounding of the greenhouse effect) has only become known much more recently. In the absence of such evidence it was at least understandable that conservationists like Wallis continued to think in 1970 that the real problems were for the most part local and, though serious, could be managed in a piecemeal way.

3.6   The growth of Green politics

Since the 1970s environmental considerations began to occupy an increasingly important place on the political agendas of European countries. The story is complex and varies from one country to another. Geographical location makes obvious differences to what are perceived as the main problems. Acid rain was a major issue in Norway, on the lee side of the polluting source, when few people in Britain knew what it was. The Dutch, at the mouth of the Rhine, have been worried about river pollution and, with so much low-lying country, about the threat of a rise in sea-levels. And so on.

In some countries, such as West Germany, an environmentalist political party (Die Grünen) began to make a considerable impression by the late 1970s and early 1980s. In 1978 a poll showed that 38% of voters between the ages of 18 and 34 supported the Greens in the state elections in Hamburg. By 1983 the German Greens had polled more than 5% of the popular vote across the country. According to the West German system of proportional representation, this entitled Die Grünen to send representatives (27 of them, as it turned out) for the first time to the Bundestag in Bonn. In Britain, by contrast, a lost deposit was the usual result for Green candidates, at least up until 1989, when they improved their vote significantly in the elections to the European Parliament. However, since then levels of support have fallen.

The progress of Green parties is not by itself an accurate barometer of increased concern about environmental issues. In the Netherlands, for instance, it has been claimed that the reason why there was no Green party is that the main political parties put a considerable emphasis on environmental questions, in response to public pressure. The relative success of Die Grünen has been partly due to its strong commitment to nuclear disarmament at a time when the major political parties in Germany supported the retention of NATO nuclear bases on German soil.

Environmental issues have played an important role in the politics of Central and Eastern Europe. Since the late 1940s rapid industrialisation created serious pollution problems in many parts of the Soviet bloc. For most of the period, no criticism or protest was allowed. After 1977, environmental protest began to grow in many countries as this was a form of dissent which could be expressed as 'for the common good' and without overt criticism of government. Green issues and groups were extremely active in the fall of communist governments in 1989 and Green parties were expected to do well thereafter. However, few Green MPs were elected in the first free elections, and as the new governments became preoccupied with employment and economic growth, environmental issues have been given lower priority. By the early 1990s pressure from the European Union had become the main factor encouraging the new governments to take action against pollution (Waller & Millard, 1992).

3.7   Review

In this section we have been looking at some of the evidence of a new kind of concern about environmental questions that began to be widely shared by around 1970 and to which the word 'environmentalism' has come to be attached. The evidence of environmentalist organisations, although relevant, gives the impression that environmentalism is a more radical phenomenon than it was. This picture is corrected by looking at the more conservative environmentalism embodied in instruments of public education such as encyclopaedias. The environmental movement thus emerged as a 'broad church', embracing a wide range of views and encouraging widely differing kinds of activity and lifestyle. In this section the emphasis none the less has been on what is common to the new environmentalism of the early 1970s, in particular the sense of humanly induced global environmental crisis. Its diversity, and its links with various traditional attitudes to Nature, is more the concern of the next section.

4   Varieties of environmentalism

One of the results of the sudden upsurge of interest in environmentalism around 1970 was a baffling variety of different interpretations and emphases. This prompted a need to classify the approaches in order to identify the main categories of environmentalism. Many of the classifications were simple dualisms: Section 3 suggested differences between 'establishment and radical' environmentalism; the media often used 'shallow versus deep green'. An influential classification in the academic literature was O'Riordan's (1981) division into 'technocentric vs ecocentric' – terms which were described by Pepper (1984) as follows:

> **Ecocentrism**   Ecocentrics lack faith in modern large-scale technology and the technical and bureaucratic élites, and they abhor centralisation and materialism. If politically to the right they may emphasise the ideas of limits, advocating compulsory restraints on human breeding, levels of resource consumption and access to nature's 'commons'. If to the left, their emphasis may be more on decentralised, democratic, small-scale communities using 'soft' technology and renewable energy, 'acting locally and thinking globally'.
>
> **Technocentrism**   A 'mode of thought' which recognises environmental 'problems' but believes either unrestrainedly that man will always solve them and achieve unlimited growth (the 'cornucopians') or more cautiously, that by careful economic and environmental management they can be negotiated (the 'accommodators'). In either case considerable faith is placed in the ability and usefulness of classical science, technology, conventional economic reasoning (e.g. cost–benefit analysis), and their practitioners.   (Pepper, 1984, pp. 237, 241)

Although widely quoted and useful in connecting attitudes to policy responses, the dichotomy seems at best incomplete, as it has no place for Romanticism, which is neither technocentric nor ecocentric, nor for

Utilitarianism, which aims to consider all sentient beings. It has also been seen as flawed in its emphasis on technology as the opponent of the environment rather than the **anthropocentric** (human-centred) use of the technology. A more recent trend has been to see the central dualism as between anthropocentrism and ecocentrism and then to recognise intermediate positions along that continuum.

One of the most systematic discussions of recent environmentalist value positions is Eckersley (1992). She argues that the new element of recent years is the growing acceptance of a degree of ecocentrism, and a move from a concern with the distribution of environmental costs and benefits to debates about new ways of life which would emancipate people from the distortions of industrial society, as well as give the environment some priority over purely human demands.

She sees actual and possible environmentalist positions as varying in the degree to which they admit ecocentric values, and discusses them under five headings on a spectrum from anthropocentric to ecocentric, as follows.

1   *Resource conservation*   Although she identifies early advocates of this position, including Plato, she gives Gifford Pinchot the central role. In this view, conservation is the professional management of development for the benefit of the many rather than the profit of a few. Development seeks to eliminate waste – including the waste involved in leaving natural resources unused. This position is explicitly Utilitarian and accords no intrinsic value to the non-human world.

2   *Human welfare ecology*   This variety of environmentalism is centred on a concern for a safe and pleasant living environment for people. Its roots go back to the campaigns to improve nineteenth-century cities, notably the movement led by Edwin Chadwick in the UK. It is concerned not only with production, but also with ensuring good conditions for reproduction – both in the sense of having healthy children and of raising them to become producers and parents, and hence reproducing society over the generations. It has been strongest in Europe, often in the form of middle-class campaigns against air and water pollution in urban areas. It was the main impetus behind Die Grünen. It remains Utilitarian and anthropocentric.

3   *Preservationism*   This approach, forged largely in the wild places of the New World and with John Muir as the early leader, is the first of Eckersley's categories which recognises Nature as being valuable in its own right. However, its focus on aesthetically appealing landscapes makes it very selective, and the emphasis on wilderness tends to emphasise a separation between society and Nature which may not exist even in remote places. For example, Yosemite valley was managed for centuries by the Awaneechee and even extremely remote areas like Alaska have been settled for millennia (Sarre, 1995).

Wilderness preservation has been quite successful in practice, but often because of its appeal to human interests, whether for aesthetic, spiritual, scientific or recreational purposes. Although its ambitions are limited, it does represent the most successful environmentalist challenge so far to the focus of industrial society on material production and consumption. Some of the arguments for preservation draw on the idea of Stewardship, but the strongest advocates are influenced by Romanticism.

4   *Animal liberation*   Eckersley includes this position in her spectrum because it is an important step away from anthropocentrism, but she indicates that in some respects it is not really an environmentalist position

at all. She traces the practical origins back to the societies for the prevention of cruelty to animals and the theoretical origins to Bentham's Utilitarianism. The leading modern advocate is Peter Singer, who emphasises the interests of 'sentient beings'. He identifies a problem in defining which animals feel pain and which do not, and concludes that the line should be drawn 'somewhere between a shrimp and an oyster'. This definition of sentience seems based more on convenience than on science.

Animal liberation is also open to question from an ecocentric position as it accords consideration to 'sentient' animals but not to the plants which provide the basis of their food supply. Its emphasis on the preservation of individuals ignores the organisation of ecosystems and the rarity or abundance which arguably makes some individuals (e.g. wild pandas) more important than others (Australian rabbits or sewer rats). In the last analysis, as Nash (1989) has shown, the extension of rights from kings to barons and then to aristocrats, property owners, working men, slaves, women and sentient animals is an historic process which may continue to extend towards less sentient animals, plants, ecosystems, streams, mountains and even rocks. It is by no means clear that Singer's cut-off point, based on a debatable concept of sentience, is any more justifiable as a principle than any other.

5   *Ecocentrism*

An ecocentric perspective may be defended as offering a more encompassing approach than any of those so far examined in that it (i) recognises the full range of human interests in the nonhuman world (i.e. it incorporates yet goes beyond the resource conservation and human welfare ecology perspectives); (ii) recognises the interests of the nonhuman community (yet goes beyond the early preservationist perspective); (iii) recognises the interests of future generations of humans and nonhumans; and (iv) adopts a holistic rather than an atomistic perspective (contra the animal liberation perspective) insofar as it values populations, species, ecosystems, and the ecosphere as well as individual organisms.   (Eckersley, 1992, p. 46)

The central feature of ecocentrism is a rejection of the anthropocentrism which has characterised Western thought for two millennia and especially the modernist world view. Ecocentrics have a central belief in the relatedness of individuals in Nature and society. Rather than emphasising the existence of distinct and independent individuals, they argue that individuals are constituted by their relationships. The ecocentric's approach to both society and Nature is holistic, emphasising the interactions within natural systems, within human society and between human society and its environment. They argue that modern science is much closer to this view than to the mechanical universe proposed by the modernists.

Eckersley concedes that, although the philosophical basis of ecocentrism is clear, there is a long way to go to clarify the economic and political arrangements which it implies, and even further to go to persuade current vested interests to adopt them. To summarise a long book, she argues that ecocentrism is incompatible with conservatism, liberalism and socialism but has a potential compatibility with anarchism. In so doing, she is defining an incompatibility between ecocentrism and the three political traditions which, in varying combinations, dominate politics in the industrialised societies (Box 3.2). Small wonder that environmental issues have made only a minor impact on political priorities in most parts of the world, and then through limited claims for resource conservation, social welfare ecology, wilderness preservation and animal liberation.

Box 3.2   *Political traditions and the environment*

Political scientists recognise three major traditions of political thought and action. They are described here in British terms, though similar principles can be recognised in other countries, even where political parties bear different names.

1   Conservatism is the oldest tradition. It favours stability and therefore supports the existence of a social hierarchy, including monarchy and aristocracy (or their republican equivalents) and the lower orders 'knowing their place'. It also supports community and nation, so the privileged are seen as having duties and responsibilities. The environmental attitude most compatible with this tradition is Stewardship. Indeed, the concept had legal force in the past as the basis on which aristocrats held their estates.

2   Liberalism was the political tradition launched by modernism. It stresses the rights of individuals, advocates minimal regulation by government or religion and trusts in enlightened self-interest to spur economic growth and social progress. Utilitarianism was formulated by and for liberals.

3   Socialism is the youngest of the three traditions, dating back to the early years of the Industrial Revolution. It aims to provide equally for all and emphasises the role of the state. In Marxism the state acts as owner of the means of production; in social democracy the state acts through taxation, regulation and the provision of social services. Social welfare ecology is the environmental attitude most closely linked with this tradition.

These traditions change over time and can be mixed even within a single party – for example, Thatcherism combined a liberal free market view of the economy with an authoritarian conservative emphasis on social order and the right to manage.

The one feature that connects all three traditions is that they are anthropocentric. Attempts to develop ecocentric political parties, usually known as Green or Ecology parties, are recent and have attracted only minority support at elections. As a result, environmental politics is mostly extra-Parliamentary, acting through pressure groups in association with parallel struggles like feminism or as direct action against particular projects. Opposition groups, of which the anti-road protest movement is currently pre-eminent (1995), can bring together people from widely different traditions, but so far these alliances seem to be only temporary.

---

5   *Environmentalism at the millennium*

The previous sections suggested that environmentalism has made relatively minor gains in terms of real world politics, and then largely in its more anthropocentric variants. A more ecocentric position has been explored but this is philosophically, politically and economically radical and has therefore been unattractive to practical decision-makers. At present it seems likely that this situation will remain in place indefinitely and that environmentalists will continue to act as critics, able to achieve some modifications to policies defined by the need for corporate profit or political expedience, but unable to achieve any more fundamental shift in values.

One line of argument throws doubt on this 'business as usual' scenario. Simmons (Chapter 2) has argued that shifts in forms of production and consumption have dramatically changed ways of life. Earlier sections of this chapter have argued that some of those shifts have been followed by major changes in attitudes to Nature. This becomes relevant to the current state of environmentalism because many social theorists believe that society has been experiencing another great change at the end of the twentieth century. The modest, and probably inconvertible, formulation is that this process is one of globalisation, not only of the economy but of politics and in some respects even culture. If environmentalism is to exert any influence in this rapidly changing world, it too needs to tackle global concerns. Two recent contributions to environmentalist thought, examined below, do promise to make the debate global, although they do so in very different ways.

5.1   Sustainable development

At the time of writing, the most sustained effort to promote global change to solve environmental problems has come from the United Nations Commission on Environment and Development. It generated the Brundtland Report (1987) and the Rio Earth Summit in June 1992.

The Brundtland Report was the culmination of a three-year process of consultation, analysis and negotiation. It tackled two massive problems – development and environment – and claimed to offer solutions to both. It demands attention because it has been so influential, but more than a decade after its publication the problems have worsened and the response is inadequate, so the attention paid to it must be critical, seeking the limitations as well as the achievements.

A central part of the Report's argument concerns the problems generated by the 'breakdown of compartments' between nations and between sectors and the creation of an interlocking global crisis of population, environment, development and energy. Rather than being daunted by the complexity of the crisis, the Commission adopted an idea first publicised in the World Conservation Strategy of 1983 – a new form of development, to be known as **sustainable development**, which would *'meet the needs of the present without compromising the ability of future generations to meet their own needs'*.

The focus on needs has the strength that it puts the interests of the less developed world high in its priorities – an understandable position since a majority of the Commission's membership came from the less developed world. This is clearly positive in principle, but it could be seen as a weakness in practical terms, since although this issue had been emphasised a few years earlier by the Brandt Report, little had been done to change the economic and political processes which lead to uneven development. However, there is also a serious weakness in principle: the needs being addressed are human needs and, in spite of the 'one earth' in the title of the Commission's 'Overview' and their recognition of 'spaceship earth', Nature is regarded only as a resource base to be used wisely. The stress on greater equity makes the report humanist in the best sense, but the neglect of Nature's interests makes it anthropocentric in the derogatory sense.

The main body of the Report consisted of six chapters dealing with specific topics followed by three dealing with key areas of policy. The six topics were population and human resources, food security, species and ecosystem conservation, energy, industry and urbanisation. The analysis suggested that the six topics were interlinked and that all posed issues in the policy areas of managing the commons, peace and security, and institutional change.

The Report's concluding chapter is a call for massive institutional and legal change to promote sustainable development policies on all scales from local to global. The problem here is motivation: such changes could be made if all governments accepted the need, but how can the governments and citizens of the first world be persuaded to prioritise the present needs of the less developed world and the needs of future generations over their established goals of higher consumption in the immediate future? The nearest approach offered by the Commission is in the chapter on 'peace and security', where conflicts over migration and resources are predicted to increase if sustainable development is not achieved. However, the potential problems for the first world are not stressed as much as the potential danger from nuclear war or the cost of global spending on arms and military forces. The Report points out that 'environmental stress is both a cause and an effect of political tension and military conflict' (p. 290), but the large majority of the conflicts cited (including a short paragraph on water) are in the less developed world, leaving the impression that the developed countries need not be involved. Perhaps most puzzling is that the chapter on 'managing the commons' covers the oceans, Antarctica and space but omits the most obvious common resource – the atmosphere.

It may be that the Brundtland Commission was unfortunate in that most of its work was done in the mid-1980s, before ozone depletion and global warming had emerged as major issues. However, both issues are mentioned sporadically in the Report without ever coming centre stage, so it is more likely that the pervasive anthropocentrism of the Report marginalised these problems. Paradoxically, in subsequent years public concerns about 'the ozone hole' and the effects of global warming were much more extreme than concern about the current devastation of lives and environments by poverty. Indeed, in the late 1980s, action against ozone depletion was more rapid and determined than action over any other environmental problem before or since. When the Brundtland process reached its climax at the Earth Summit in Rio de Janeiro in 1992, most of its concerns were expressed only as desirable directions for policy in Agenda 21. Two issues generated sufficient support to be put forward as conventions for signature – the Climate Convention, tackling emission of greenhouse gases, and the Biodiversity Convention, focusing on conserving genetic resources. Problems which involve the large-scale destabilisation of environmental systems seem to stimulate greater concern than human problems, even when their consequences for humans are uncertain. Most environmentalists are very critical of what was achieved at Rio, but more was done for certain global environmental issues than for development.

Today, the Brundtland Report and the concept of sustainable development are seen in profoundly ambiguous ways. On the one hand they have stimulated the most significant political response yet to environmental issues. Most world governments have adopted the language of sustainability and are involved in on-going policy debates. On the other hand, environmentalists criticise the anthropocentrism and the degree of vagueness which makes it possible to espouse the concept without any significant change to priorities or practices. Most politicians assume that the degree of change advocated by the report is far too radical to be contemplated, let alone implemented. Environmentalists are beginning to argue that not only is sustainability virtually impossible to define in practical terms but also it is in principle the *minimum* environmental standard anyone could possibly advocate. To opt explicitly for non-sustainable treatment of resources and environments is to suggest handing on a poisoned chalice to future generations. No one would advocate this as desirable, though this is precisely what global society is doing in practice.

In these circumstances there seem to be two options for environmentalists: put more energy and ingenuity into promoting sustainable development, or find new ways of reconciling environmental goals with those of citizens in all parts of the world.

5.2   The Gaia hypothesis

The most thought-provoking exponent of a truly *global* representation of the environment is James Lovelock (1979), through his Gaia hypothesis. Lovelock is an unconventional scientist who was inspired to formulate the idea of Gaia while working for NASA as a contributor to the design of a lander to seek for traces of life on Mars. He realised that the obvious indicator of life on Earth is the composition of the atmosphere: a large proportion of a highly reactive gas like oxygen is dependent on living things to offset the chemical processes which would naturally capture the oxygen. From this insight, he has developed not only a holistic view of geology and biology (much of which is acceptable to other scientists) but also the suggestion that the whole system is a self-regulating being, named Gaia after the Greek earth-goddess. Lovelock denies any supernatural or teleological intent and insists that he regards Gaia as a testable hypothesis, but his view has been more attractive to New Age environmentalists than to most scientists. However, as more work is done on global warming and other problems of global linkages, more scientists are beginning to accept the existence of natural feedback processes which act to regulate the state of the atmosphere and biosphere. Arguably, it is a most significant development as it integrates all environmental processes and situates human society in Nature.

Lovelock's account of the development of Gaia suggests that the evolution of living things progressively changed the atmosphere, and with it the temperature of the Earth's surface in ways that allowed more complex life forms and ecosystems to evolve. Lovelock shows that this process effectively insulated living things on the surface from cosmic rays and fluctuations of solar radiation through the structure and properties of the atmosphere. A parallel argument relates the Earth, Mars and Venus to the Goldilocks story: from rather similar planetary beginnings, the Earth and its atmosphere has ended up 'just right' for life, Mars is too cold and Venus too hot. It could be that it is just tautology: we would not be here to marvel at the improbability of the Earth's evolution if we had not been part of it. However, for many people Gaia offers a creation story that both excites a sense of wonder and seems scientifically credible, though it is far from being scientifically proven.

By itself, the idea of Gaia does not offer detailed policy prescriptions, but it does suggest a significant reorientation. Lovelock's hypothesis is planet centred and not human centred, and rather than suggesting that the human species is the culmination of evolution, it tends to suggest that we may be an aberration. From a planetary point of view we are seen more as a runaway disease, growing rapidly at the expense of other organisms and disrupting essential natural cycles. Fortunately for the planet, Lovelock recognises great resilience in the face of such disruptive events (including recovery from the huge comet impact which is now seen as the likely cause of the extinction of the dinosaurs). Unfortunately for the human species, he suggests that if we are too disruptive this resilience will allow the planet to eliminate us.

The Gaia hypothesis dramatically contradicts the modernist idea of science as analytical, atomist, reductionist and mechanical. It builds on earlier thinkers like pioneer geologist James Hutton (1726–97), who argued that processes like mountain building and erosion were necessary to produce the soils on which plants and animals depend, and Charles Darwin (1809–82), who explained how natural selection could have guided evolution towards the complex organisms and ecosystems of the natural world, in order to portray a world which is highly integrated, changing over time, and adaptable to changing circumstances. It is an image of the world which excites admiration and even awe – a globalisation of Romanticism.

Lovelock's views seem to leave a range of choice for humans: we can focus on scientific testing of hypothesised feedback links; we can behave self-indulgently and leave it to Gaia to punish us if she can; or we can assume that (as 'Nature grown self-conscious') we should act as stewards for the whole process of planetary resilience and species evolution. The third option implies that we should be thinking of handing the Earth in good order to our descendants, not in the shape of our children but in terms of our evolutionary successors. To do that, we would need to give a high priority to the maintenance of biodiversity, not frozen in gene banks or zoos, but living free in varied ecosystems.

The Gaia hypothesis is extremely challenging both because it is profoundly ecocentric and because it is on such a large scale as a scientific hypothesis. The ecocentrism challenges our assumption that the human species is central to life on Earth. The scale makes it difficult to verify or falsify it, as could be done with a more modest hypothesis. It is likely that debates between Gaians and more conventional scientists will remain unresolved, even though all sides are much more aware of complex feedback processes between living things, atmosphere, rocks and solar radiation. Meanwhile, Gaia does not need to be scientifically validated to act as a stimulus to recognise the age, complexity and beauty of the natural world and to regulate our use of it to sustain, or even enhance, its admirable features.

6   Conclusion

This chapter has covered a huge span of time from hunter–gatherers to the electronic age, and including significant transitions around ten thousand, one thousand, one hundred and, perhaps, ten years ago. It has also covered a huge span of thought, from cultures which were immersed in Nature, through cultures which sought to separate themselves, practically and conceptually, from Nature, to what may be the dawn of a new scientific understanding of the evolution of the cosmos, life and the human species. Our survey shows that attitudes to environment have varied a great deal over the period. In conclusion we want to highlight three points, two negative and one positive.

First, the exploitative attitudes which dominate modern society's practical use of environment seem ultimately to stem from philosophical, religious and scientific ideas which no longer command wide support. The Ancient Greek pre-eminence of male rationality, the Old Testament injunctions to multiply and have dominion over the Earth, and Bacon's mechanistic universe have few conscious adherents but seem still to exert an effect on our attitudes. These attitudes are not well articulated – whereas most 'primitive' societies have a clear view of their place in Nature and their entitlement to use natural resources, in practice modern society seems to prioritise 'growth' and 'material consumption' even when claiming to support Stewardship, sustainability or ecocentrism.

Secondly, and more positively, the last generation has seen a great upsurge in concern about environmental problems, including pollution, resource exploitation, loss of biodiversity and atmospheric change. As yet, this is a heterogeneous and fragmentary movement and its successes are limited to mitigating the impact of development. Yet there are some grounds for optimism: the rapid globalisation of the economy, politics and communications has brought environmental issues into international politics where, for instance, polluting industries may be opposed by governments rather than just NIMBY protest groups. Increasingly, protest groups can use electronic communications to spread ideas, seek information and seek international support. The slogan 'think global' is being developed into a truly global understanding of the biosphere as a whole system – an understanding which is being expressed by scientists through predictive computer models and by New Age mystics through crafts and ceremonies. The potential seems to exist for a new world view which links a scientific understanding of our place in Nature to social beliefs which define the goals of life in ways which are not environmentally damaging.

However, the third and final point is more pessimistic. History suggests that institutions like governments and corporations are much more concerned with social considerations than environmental factors. Among those concerns, political and economic power are much more frequently exercised to achieve or resist dominance than for altruistic purposes. Paradoxically, the societies where altruism has been taken most seriously are the liberal democracies, inspired by just those humanistic ideals which have been associated with increased environmental exploitation. Present-day world society, through increasingly integrated, remains fragmented into nearly two hundred nation states with different organisation, power and interests, so any move toward more environmentally friendly politics faces huge obstacles.

To sum up, the most likely prospect is that environmentalism will only achieve marginal gains through the variants (resource conservation, human welfare ecology) that are more compatible with existing conservative, liberal and socialist regimes through their appeal to Stewardship or Utilitarian arguments. Brundtland's concept of sustainable development adopts precisely this combination. More ecocentric forms of environmentalism remain improbable without a fundamental shift in social priorities, although a case can be made that they could be socially as well as environmentally preferable. Whether pursuing practical gains or a more ambitious reorientation, environmentalists face the problem of reconciling social practices with developing scientific understanding and technical possibilities. Past history suggests that rationality will play a part in that reconciliation but that social and environmental values will play key roles in setting priorities.

Activity 5

Consider further an environmental issue in your own locality. What are the attitudes of the interest groups involved? Do you think their differences could be reconciled? If so, how? If not, with which view are you inclined to agree?

References

ALLABY, M. (ed.) (1988) *Macmillan Dictionary of the Environment*, Macmillan, London, first published in 1977.

BACON, F. (1620) *Novum Organum*, London.

BULLOCK, A. & STALLYBRAN, O. (eds) (1977) *The Fontana Dictionary of Modern Thought*, Fontana Books, London.

CARLYLE, T. (1829) The signs of the times, *Edinburgh Review*.

CARSON, R. (1965) *Silent Spring*, Penguin Books, Harmondsworth, first published 1962.

ECKERSLEY, R. (1992) *Environmentalism and Political Theory*, UCL Press, London.

GRIGGS, E. L. (ed.) (1956–71) *Collected Letters of Samuel Taylor Coleridge*, Clarendon Press, Oxford.

LOVELOCK, J. (1979) *Gaia: a new look at life on Earth*, Oxford University Press, Oxford.

MALINS, E. (1966) *English Landscaping and Literature 1660–1840*, Oxford University Press, London.

MERCHANT, C. (1980) *The Death of Nature: women, ecology, and the scientific revolution*, Harper & Row, New York.

NASH, R. (1989) *The Rights of Nature: a history of environmental ethics*, University of Wisconsin Press.

NICHOLSON, E. M. (1970) *The Environmental Revolution: a guide for the new masters of the world*, Hodder & Stoughton, London.

OELSCHLAEGER, M. (1991) *The Idea of Wilderness: from prehistory to the age of ecology*, Yale University Press, New Haven.

O'RIORDAN, T. (1981) *Environmentalism*, Pion, London.

PASSMORE, J. (1978) *Man's Responsibility for Nature: ecological problems and western traditions*, Duckworth, London.

PEPPER, D. (1984) *The Roots of Modern Environmentalism*, Croom Helm, London.

PLUMWOOD, V. (1993) *Feminism and the Mastery of Nature*, Routledge, London.

PORRITT, J. (ed.) (1987) *Friends of the Earth Handbook*, Macdonald, London.

SARRE, P. (1995) Paradise lost, or the conquest of the wilderness, in Sarre, P. & Blunden, J. (eds) *An Overcrowded World?*, Oxford University Press, Oxford.

SARRE, P., SMITH, P. & MORRIS, E. (1991) *One World for One Earth: saving the environment*, Earthscan Publications in association with the Open University and the World Wildlife Fund for Nature, London.

STUDENTS YEAR BOOK (1970) (covering the year 1969) Annual Supplement to *Collier's Encyclopedia and Merit Students Encyclopedia*, Crowell-Collier Educational Corporation, New York.

THOMAS, K. (1983) *Man and the Natural World: changing attitudes in England 1500–1800*, Allen Lane, London.

WALLER, M. & MILLARD, F. (1992) Environmental politics in Eastern Europe, *Environmental Politics*, **1(2)**, pp. 159–185.

WALLIS, H. F. (1972) *The New Battle of Britain: a conservation handbook and directory*, Charles Knight, London.

WHITE, L. (1967) The historical roots of our ecological crisis, *Science*, March 10, pp. 1203–7.

WHITE PAPER (1970) *The Reorganisation of Local Government*, HMSO, London.

WORLD COMMISSION ON ENVIRONMENT AND DEVELOPMENT (1987) *Our Common Future* (The Brundtland Report), Oxford University Press, Oxford and New York.

Answers to Activities

Activity 2

We can review the four perspectives in relation to the question of why the loss of a species of flower might matter:

1   The Stewardship perspective sees each species as having a place in a divinely ordained scheme of things. According to this view there is something inherently bad about a species being lost.

2   The Imperialist would not think there is anything inherently bad about a species being lost. But he would be a fool not to listen to the point made at the beginning of Chapter 6 that it is not rational to use a treasure-chest for firewood without looking carefully to make sure there is no treasure inside it. Since many tropical forest plants have been found to be very beneficial to the human race – for instance, by providing medicines that have not been synthesised in the laboratory – there is potentially a considerable long-term loss in allowing unstudied or understudied plant species to disappear. There might be a disagreement between people who shared an Imperialist perspective as to whether the risk of losing what might have been an important medicine should be taken. In practice, however, it is likely that an Imperialist will prefer the more certain short-term economic gain from destroying a habitat to the less certain long-term gain that might come from preserving it.

3   There may be more than one Romantic response. A Romantic might want to keep the tropical forest in its original, wild state. This would mean preserving the species in the forest. But this may be incidental. Coleridge was fascinated by desolate places where he believed (wrongly) nothing lived. On the other hand, a pantheist – and Romantics were often pantheists – ought to believe that all things are sacred and hence that it would be a desecration to destroy the plants. Their rarity or potential usefulness would not come into it.

4   The Utilitarian, like the Imperialist, might in theory come down on one side or the other in this particular case (see under 2).

Activity 4

Some differences may seem so significant as to overshadow any similarity between the radical environmentalism of bodies like the Friends of the Earth and the environmentalism of the encyclopaedia contributor. The author of the *Students Year Book* article is a kind of civil servant and so, to some extent, expresses establishment attitudes. He expects initiatives in environmental matters to come from educators, resource managers and environmental scientists, not from 'the man in the street'. Jonathon Porritt, by contrast, writes of the large numbers of ordinary people involved in the activities of the Friends of the Earth. Such grassroots organisations will engage in 'direct' action of a kind that we can readily imagine government officials would not approve of. These actions may even be directed against government policy (for instance, through anti-nuclear activities) or in the face of officially authorised approval (where planning permission has been given for the development of what had been a valued wildlife site). The emergence of groups like the Friends of the Earth is by no means what the author of the *Students Year Book* article had in mind when he wrote of 'dramatic changes in the conservation movement'.

None the less there are reasons for regarding the emergence of Greenpeace and Friends of the Earth as part of the same development reported by the *Students Year Book* article. Although their mode of operation in trying to get things done is unlikely to have been acceptable to establishment environmentalists, there are fundamental points in common to their way of thinking, as can be seen from a consideration of the four new emphases to which the Year Book article refers:

1   The new 'conservation philosophy', according to the *Students Year Book*, is comprehensive where it had previously been compartmentalised. One example given is of the new willingness of environmental scientists to try to take a large view of environmental problems, seeing pollution, for instance, not just as a technical problem (for, perhaps, chemists) but as having two main roots: over-population of the world generally and the demands on the environment of affluent societies. This emphasis on a more comprehensive view is reflected in the new environmental organisations such as Friends of the Earth, who are distinguishable in just this respect from more specialised bodies such as the Royal Society for the Protection of Birds.

2   Connected with this more comprehensive view is a willingness to put human beings and their lifestyles at the centre of environmental problems. Greater affluence, for instance, means a greater demand for cars and therefore more pollution of the air. It means greater demand for energy and therefore, in practice, a greater burning of fossil fuels with consequential environmental problems such as acid rain and compounding of the greenhouse effect. It means a greater amount of waste. And so on. This awareness of human lifestyles as a central environmental problem is another feature of the new environmental organisations. Jonathon Porritt mentions cycle rallies as one activity of the Friends of the Earth, as well as protests against acid rain. Mention might also be made of their direct action on recycling and returning all those throwaway soft-drinks bottles to the manufacturer as evidence that Friends of the Earth were also seeing humans as a major cause of environmental problems.

3   The *Students Year Book* article identifies as a further feature of the new conservation philosophy that its emphasis is global rather than local. The author seems to have had in mind that the demands on the Earth's natural resources caused by over-population constituted a global problem. Both Friends of the Earth and Greenpeace, in their opposition to nuclear testing in the Pacific, and also on other issues, have been concerned with pollution on a global scale.

4   Finally, the new conservation philosophy, according to the Year Book article, is 'less rural and more urban'. Traditional conservation bodies in the United States, such as the Sierra Club, have tended to focus on the protection of wildlife and of wildernesses. Even in Britain they have tended to be rural, e.g. the Royal Society for Nature Conservation and the Council for the Protection of Rural England. The National Trust, although concerned with the conservation of buildings and places of historical interest even in urban areas, gives pride of place to its country houses and areas like the Lake District. In general, apart from certain special cases (Hampstead, central York, Edinburgh and Bath, for instance) conservation bodies have been much less concerned about the urban environment. But many problems, such as air pollution, are apt to be worse in urban environments. And these are among the kinds of problem that have concerned environmental organisations like Friends of the Earth.

Chapter 4 Environment and development

1 Introduction

In this chapter we pursue the relationship between 'environment' and 'development'. These terms are already familiar from the previous chapters, and have now acquired strong political overtones. **Development** describes all those technological changes which seek to improve human welfare, so it concerns us all. But the traditional sequence of development towards a 'technological society' with ever-increasing energy use, as outlined in Chapter 2, is now more subject to critical questioning, as we have already seen in Chapter 3. This has led many in 'developed' industrial countries like ours to be particularly concerned about environment issues, some of which impinge on the development aspirations of less industrialised countries. This results in tension in international debate, as we shall see. The idea of 'sustainable development' is one attempt to reconcile these differences, though its meaning and realisation remain problematic. Here we shall introduce some of the general intellectual tools available for investigating these issues.

The chapter is structured around a discussion of the following questions:

- How are environment and development issues perceived globally in the 1990s?

- What processes brought us to this position?

- What constraints are implied for the future?

Two approaches will be developed in the course of the discussion which will provide frameworks for the continuing consideration of environment and development issues throughout this book, and the three other books in the series:

1   the different scales in time and space, from past to present and local to global, on which processes of change occur;

2   a descriptive formula relating population, resource use and technology.

These are followed by brief discussions of the relationships between environmental change and damage, and between development and growth.

This is only the beginning of an assessment of the progress made so far with the formidable reconsideration of international policies implied by the ideal of a sustainable global society.

As you read this chapter, look out for answers to the following key questions:
- How are environment and development issues perceived globally in the 1990s?
- What processes brought us to this position?
- What constraints are implied for the future?

2   Present perceptions

We begin, at a general descriptive level, with a discussion of our first
question:

> How are environment and development issues perceived globally in
> the 1990s?

2.1   Divergent views

As we have seen in the earlier chapters, and no doubt from our own
experience, there is no lack of evidence of environmental *change*, which
many regard as environmental *damage* or even as an environmental *crisis*.
These perceptions – change, damage, crisis – are themselves strongly
dependent on individual experience and situation. There are major
differences between the views of rich and poor, urban and rural
populations, 'developed' and 'developing' countries, 'left' and 'right' in
politics – all distinctions to be considered.

Activity 1

Before reading further, pause to reflect on your present attitude to a
particular global environmental issue: the effect of the rising human
population of the world. How do you think your attitude might differ
if you lived in China; if you were a subsistence farmer; if you were a
member of the Institute of Directors or of Greenpeace?

The growing total human population – made possible by the agricultural,
trading, industrial and financial systems developed over the last few
thousand years – has created effects on its surroundings which are no
longer purely local, as they have been in the past, but are increasingly
global in character. They form an interlocking web of issues which are often
referred to piecemeal in the media, and are not easy to disentangle or to
visualise as a whole: destruction of tropical rainforests, species and habitat
loss, overfishing, soil degradation, desertification and erosion, oil and
mineral resource depletion, air, land and water pollution, the ozone 'hole',
climate change. Such environmental change has accompanied the
'globalisation' of technology, economics, and perhaps of culture, which is a
distinctive feature of the present, not to be found in earlier human history.
The separate national institutions (government, financial and legal
systems), evolved worldwide to deal with local problems, are no longer
able to deal with these larger-scale issues, and some international regimes
are beginning to be constructed to provide a more global system, as we
shall see. However, political conflicts between sovereign nation states
continue to dominate such international discussion.

Different countries, which used to be crudely classified into First
(market), Second (centrally planned) and Third (uncommitted) World, are
now increasingly, and equally crudely, seen as 'rich' industrialised North

versus 'poor' debt-ridden South. This distinction is reflected in the 'G7' group of industrialised countries (originally USA, Japan, West Germany, France, UK, Italy, Canada, but now including the EU) which seek to control international trading arrangements, and the 'G77' group of other nations – in fact now about 130 countries – which often form a loose coalition of opposition to them in international debate. There are political value judgements implied in all these terms, which fail to recognise major differences between countries within the categories, and between different groups of people within each country. The rapidly expanding industrial economies in South-East Asia, or the oil-rich countries of the Middle East, don't fit simply into this classification; Australia and New Zealand are regarded as belonging to the 'North'; established industrial countries like the UK and USA have significant problems of poverty and unemployment; the former Communist countries of Central and Eastern Europe are more pre-occupied with economic reorganisation and local industrial pollution than with global questions. All these differences lead to very varied perceptions of the relative importance of environment and development, and of appropriate action, as emerged for example at the Rio Earth Summit, the UN Conference on Environment and Development, in 1992.

Over the last decade, particularly since the 'Brundtland' report *Our Common Future* of 1987, the policy objective of sustainable development has been generally endorsed by many governments and international institutions. While this was defined as 'development which would meet the needs of the present without compromising the ability of future generations to meet their own needs', what exactly this means is not straightforward – as was seen in Chapter 3. There are divergent views between those who interpret its implications in terms of improved technology, of greater ecological awareness or of political and economic change – no doubt all have some part to play. Furthermore, the links between local activities and global consequences are not always obvious, nor are the local responses that will be required to change the global effects. Certainly the changes from current practice implied are far more radical than most governments, or their electorates, have so far been prepared to put into practice.

Activity 2

Our society, and increasingly that of all developing countries, is very dependent on road transport, with oil as its fuel. There is growing congestion, the available amount of oil is limited, and there are significant pollution effects associated with its use, both local and global (carbon dioxide emissions leading to the threat of climate change). So its expanding use is clearly not sustainable. Consider the possibilities (technical or social) for a systematic reduction in the use of oil in road transport, and their implications – internationally, nationally, locally and personally.

Activity 2 introduces just one example of the many topics where global implications may force changes in local behaviour, if they are appreciated in time. Other such topics will become clearer in later chapters and books in the series. For the present let us continue to gain a general impression of the global issues considered important at a particular moment in the early 1990s, and divergent attitudes to them worldwide.

2.2   15 June 1992

On this date, already receding into history, the United Nations Conference on Environment and Development (UNCED) in Rio de Janeiro ended. This was only the second such 'Earth Summit' ever held, following the first UN Conference on the Human Environment held in Stockholm in 1972. The addition of 'Development' to the title reflects a change in political attitudes over the intervening twenty years. Most of the 178 national governments represented in Rio signed five separate agreements which indicate the topics discussed (see Box 4.1).

What all these mean, and the general progress of international debate of which this meeting was a part, is considered in more detail in *Blackmore & Reddish (1996)*. Here we will simply try to capture a general impression of world responses to this particular global event.

Box 4.1   The agreements signed at the Rio Earth Summit, 1992

Two legally binding Conventions, on Climate Change and Biological Diversity, represented the culmination of much earlier discussion. Although disappointing in their details to those seeking firmer agreements, they do acknowledge the need for some action to reduce the risk of climate change and to reduce species loss. An important element was the setting up of systematic monitoring and negotiating procedures for extending the agreements. These are continuing to operate. Less binding was the 27-point Declaration on Environment and Development, a hard-fought compromise on all the issues dividing North and South, rather than the ringing Earth Charter that had been hoped for by the UNCED organisers. Similarly the 40-chapter Agenda 21 action plan for achieving sustainable development was also only an agreed collection of proposals, not legally binding; its limited success lies in the acceptance of sustainable development as an international policy objective only five years after Brundtland, and in providing a reference document for national and local planning. (There is a continuing Local Agenda 21 activity worldwide.) However, an acknowledged failure of Rio was the lack of agreement on the vast transfer of funds thought to be needed to implement it. Least satisfactory of all was the Statement on Principles of Forest Management which fell far short of the legally binding convention that was originally proposed.

For more details see, for example, Grubb *et al.* (1993).

2.3   World news reports

A 'snapshot' of world attitudes to Environment and Development at a particular moment is provided by comparing the coverage given to the concluding ceremonies of UNCED by newspapers from around the world on 15 June 1992 or thereabouts.

Like all snapshots, this can only give a partial picture. It doesn't show the long period of preparatory negotiations before the conference, earlier coverage of the conference itself, or the developing international institutions since. More seriously, the image presented by newspapers is distorted, by all kinds of commercial and political factors, including government censorship in many places. Other sources of information like television and radio may be more important now for many people. Nevertheless, the newspaper files are available to indicate the importance attached to this event by particular journalists on that day as contrasted with the other more local immediate topics that fill the papers. As long as they are viewed sceptically, they do give an immediate first impression of the diverse way a major global topic was seen in different places at one time, and the resulting contribution to formation of popular opinion.

Figure 4.1   Headlines from newspapers around the world on 15 June 1992.

It is immediately clear that the news coverage of UNCED had much in common worldwide – after all, everyone uses the same international press agencies. A photograph of the Cuban Fidel Castro *not* shaking hands with US President George Bush – a 'human interest' story – was used widely: the UK *Daily Express* had no other mention of the event whatsoever. A Greenpeace demonstration in Trafalgar Square, with a banner saying 'UNCED – Words failed us!' hanging from Nelson's Column, was illustrated as far away as the Tokyo *Japan Times*. Furthermore, the proportion of space devoted to the whole event was never more than a few per cent at most. Many newspapers have 40 or more pages in several sections, with five or more pages each devoted to home news, sport, business, entertainments, advertisements and magazine-type features – and it was rare to find as much as three pages in all on UNCED. It competed, in general, with news about fighting in Sarajevo and about English football fans rioting in Malmö, and more specifically with national news items. Predictably, far more effort goes into conducting and recording our normal activities than into proposals, however partial and imperfect, to change them. But in favourable cases it *was* the front page story; there was a full account of the agreements, comment from the paper's own reporters there (among the 8479 accredited journalists said to be present), independent discussion, perhaps a leading article. The interest lies in the differences: how much attention was given, what other stories mattered more, how did assessment vary?

At worst it could be very negative – the UK *Sun* had no news items at all (with the front page and six more devoted to Princess Diana) but a characteristic leader (Figure 4.2).

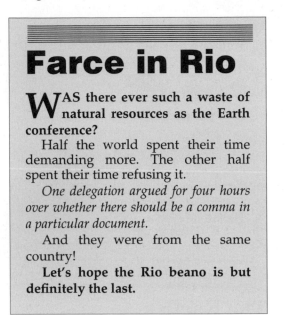

Farce in Rio

WAS there ever such a waste of natural resources as the Earth conference?

Half the world spent their time demanding more. The other half spent their time refusing it.

One delegation argued for four hours over whether there should be a comma in a particular document.

And they were from the same country!

Let's hope the Rio beano is but definitely the last.

▲   *Figure 4.2   Extract from the* Sun.

A milder version of the same sentiment was to be found from the other side of the world in a cartoon in the Australian *Canberra Times* (Figure 4.3).

▲ *Figure 4.3   Cartoon from the* Canberra Times.

However, it did have a full page of hard news, reflecting its geographical location. It particularly picked up the Malaysian Prime Minister's 'offer' to 'give back its jungle areas to the native people of Sarawak when immigrants in Australia and the United States went back to Europe'. It also noted the Australian Environment Minister's view that 'protocol took over from sense' when the Japanese Prime Minister was not allowed to address the conference by satellite in the concluding ceremonies – he was kept at home by a controversial new bill allowing Japanese troops to be used abroad for the first time since the war in UN peace-keeping missions, the *Japan Times* main story.

The concluding short speeches by 105 heads of state ('Seven Minutes to Save the World' in the *Buenos Aires Herald*) were opportunities for political rhetoric. Bush was held to be defensive, and Castro inspiring; John Major emphasised the importance of the new Sustainable Development Commission to monitor progress, but India and Malaysia saw this as 'ecological imperialism'; Germany's Helmut Kohl offered increased aid; Uganda's Museveni was 'funny'; China's Li Peng was 'dull and serious', and so on. The US *Time* magazine was worried by the impression on world opinion created by the US government stance: 'On the defensive. Who's got the hardest job on the planet? It's William Riley [head of US delegation] who is supposed to explain US positions at the Earth Summit – and keep George Bush from being the bad guy'. So was the *Wall Street Journal*: 'President's clumsy handling . . . public relations disaster'. The latter's acid third leader (after long comments on Mexican trade and Polish business) was, however, almost as destructive as the *Sun*'s, and can be contrasted with those in the UK *Times*, the *Japan Times* and the *Bangkok Post* for example (Figures 4.4 to 4.7).

Asides

Travel Section

Yes, we know there was an Earth Summit in Rio over the weekend. Everyone who thinks he is anyone was there. President Bush did not shake hands with Fidel Castro. The developed nations are sending more aid to the Third World; we hope the local kleptocrats don't take more than their share. President Bush is being castigated by the assembled, domestic environmentalists and much of the media for a 'lack of leadership,' though it looks to us as if he got what he wanted. We don't have much more to say, since so far as we can see not much happened.

▲ *Figure 4.4 Extract from the* Wall Street Journal.

Earth Summit was just a beginning

ONE fear expressed by cynics prior to the opening of the Earth Summit in Rio was that so much hot air would be emitted that it was likely to make global warming worse. Fortunately, things didn't get quite that bad. The summit did, in fact, serve as yet another much-needed reminder that we must all put together or die together on this increasingly fragile planet. And, if the politics of aid, which bedevil such North-South gatherings, could have been kept in the background the summit could have achieved more. After all, how many more times does blanket aid, except for immediate humanitarian relief, have to be shown to be counter-productive, highly-addictive and ruinous to self-help and natural development. Africa has surely proven that.

In the aftermath of Rio the new UN Commission has a very important role to play.

Its success will be crucial because the real issue at Rio was always sustainable economic growth. All environmental subjects are included in these three words. Sustainable economic growth implies not only growth which is permanent but also growth which respects the necessary criteria for man's continual presence on this planet.

The East-West conflict has now ended, and conventional wisdom sees it as having been replaced by a renewed North-South conflict, but of an economic rather than political nature this time. There is a general – and genuine – wish to bridge this divide through elaborating policies which address the major concerns of both sides. The East-West conflict was concerned with ruling the world; the new North-South conflict has more to do with saving it.

The Earth Summit is over, but let us hope that its spirit lives on and will provide fresh impetus in the war against those government and individuals who would rape and poison our planet to satiate their own greed.

▲ *Figure 4.5 Extract from* The Times.

Rio's sound and fury

Plans were set in motion at Rio to create a Sustainable Development Commission (SDC) to implement 'Agenda 21,' an 800-page program of environmental initiatives much praised at the Earth Summit. The SDC is to be a U.N. body charged with implementing Agenda 21. But will the SDC command sufficient respect and exercise adequate powers to be effective?

The answer is in doubt because Agenda 21 reflects the crisis-driven concerns of ecopessimists and carries an annual price tag of $600 billion. But the Rio summit was primarily a contest between contending governments, not environmental radicals, and so this astonishing figure was in the end only a red herring.

Far more important was the decades-old struggle between North and South over money. One perhaps surprising dimension of the Rio battle was the way that it demonstrated the overt hostility of many South or 'developing world' governments to the agenda of environmentalists who normally concentrate their fire on the rich capitalist North, including Japan. The treaty on protecting forests foundered on Southern insistence that any talk of protecting the world's rain forests in Southeast Asia or Brazil or Africa infringes on national sovereignty. Southern voices maintain that such forests are not part of the global inheritance.

Japan's prominence – especially as the world's premier creditor nation – at Rio points to four crucial roles for this nation to play. First, within the bounds of sense and efficiency, Japan must help to finance the projects agreed to at the Earth Summit. Second Rio underscores a key opening for Japanese diplomacy to surmount the North-South divide. Third, Japan has the top-notch antipollution technology that can be put to use in reducing industrial pollution in any other country. Fourth, Japanese consumers can exercise unique influence on the consumption of precious resources in ways that remind everybody, North and South, that the earth is our shared inheritance.

▲   *Figure 4.6   Extract from the* Japan Times.

GROWTH AND GREENERY

THE Rio Earth summit should finally put paid to the illusion that global problems are best addressed by mega-conferences on themes so all-embracing as 'environment and development'. The justification for this ambitious undertaking was that only a global summit could cajole politicians into committing themselves to factor into each economic decision the environmental costs to this and to succeeding generations. But Rio, heavily over-sold by its United Nations organisers, attempted too much, and in the wrong way. Its 400-page action plan is a ragbag which conflates the marginally desirable with the vital, a document so heavily politicised that it barely nods to the obstacle rapid population growth presents to protecting the environment in some of the poorest countries.

Governments have indeed endorsed the principle of 'green growth' – that our future prosperity depends on sound environmental management now. But what most politicians will take away from Rio is the conviction that environmental diplomacy is a rerun of the old confrontations in the United Nations about economic backwardness and official aid targets. Much of the past fortnight has been spent rehearsing theories development to which few governments individually subscribe, and in reviving the discredited Third World *canard* that all the world's ills are due to the West's failure to allocate a fixed proportion of its wealth to Third World exchequers.

This resurrection of North-South quarrels has raised political tempers and, worse, distorted the view of the environmental bargains to be had. Protecting the environment is not a favour the poor do the rich, in return for cash. The poor suffer most already from advancing deserts, polluted water, degraded farmland and industrial waste; they will have fewer defences against the effects of global warming. The developing world has a strong case in seeking Western cash and technology to avoid the worst environment pitfalls of economic growth; the West has an obvious interest in helping the poor avoid the mistakes it once made. The poor are rich in the biological diversity the whole world needs to conserve.

But handouts are not the whole game. Many of the most sensible steps Third World countries could take for the environment would actually save them money. Cutting subsidies for energy, logging or irrigation would profit their exchequers even before counting the green gain. And in the accounting that has to be central to reconciling growth with greenery, all countries have responsibilities as well as rights. Obvious as this may seem, it was little in evidence at Rio. The West has a duty to set exemplary standards. But there again, Rio did little to advance its proclaimed goals by consistently pillorying the United States.

Provided that the negative lessons of Rio are absorbed, the summit's legacy could still be positive. An impressive number of countries have already signed two conventions, on climate change and biodiversity.

Both are ill-drafted compromises, weak on conservation and prey to interpretation as pledges of high financial transfers from the West which will not be forthcoming. But without a summit deadline, these two conventions might have taken many more years to negotiate. The next target must be to convert Rio's disappointingly weak statement of principles on the world's forests into another convention.

The British government, which rightly resisted adding to the UN's environmental bureaucracy, supported the creation of a small UN commission on sustainable development to which governments and international agencies are invited to report. Such regular follow-up could do much to translate Rio's vague promises into national policy. The quarrelling at Rio has not entirely cancelled out the benefit of concentrating politician's minds on a greener world. Constant monitoring is the way to see that they do not forget why they agreed to be there in the first place.

▲   *Figure 4.7   Extract from the* Bangkok Post.

Activity 3

Ask yourself how the above comments might reflect general national perceptions, or only those of particular interest groups within their respective countries.

A widely used set of statistics reflected the concern of environmentalists about the number of species that had become extinct (600–900), the population increase (3.3 m), the hectares of arable land desertified (200 000) and of tropical forest destroyed (534 000) during the conference. Another quoted the crimes (including 202 homicides) in Rio during the conference – 56 foreigners, including 34 attending the conference, were robbed or mugged.

Questions about money were inevitably dominant. The *Montreal Gazette* quoted approvingly Maurice Strong, the Canadian organiser of the whole event: 'It's time for money, action, Rio chief says'. He in fact felt that 'We're on a course that's leading to tragedy' and 'We haven't another 20 years' – a reference back to Stockholm, with which he had also been involved.

His secretariat's estimate that $125 billion a year would need to be transferred from North to South to implement Agenda 21 was only matched by offers like $7 billion over 5 years from Japan, $4 billion from the EU, $0.26 billion from the USA – with all their public relations implications. The tangle between new and old commitments meant that this was variously estimated by the hard-headed as only worth $1–2 billion a year in 'new' money – described by Malaysia's Mahathir as 'like nothing at all' (Singapore *Straits Times)* or 'a pittance' (Hong Kong *South China Morning Post).*

In general the Malaysian leader seems to have been recognised as a powerful and influential voice of the G77 countries of the South. As also in general Maurice Strong was widely and admiringly quoted – though not without a counterclaim ('Strong, saviour or saboteur?' *Buenos Aires Herald),* originating with Greenpeace, that as an oil millionaire he had kept some important issues, like regulation of multinational companies, off the agenda.

An interesting local twist on funding was the view that by hosting the conference, Brazil had done particularly well with new aid commitments ('Brasil consegue na Rio-92 crédito de US$ 4 bilhões', *O Estado de S. Paulo;* 'Aid: Brazil scoops the lot', *Buenos Aires Herald).* Other South American newspapers understandably gave full accounts and comment of the conference, though not without evidence of national rivalries. Brazil's practical achievement in mounting the conference was generally praised ('Brazil takes bow for streamline success', *The Times),* though with some reminders of its problems with inflation, and brutality to its street children and indigenous peoples ('Urchins and Indians left out', *Independent).*

Other parts of the world than South America, Western Europe and the Far East were patchier in the attention given to UNCED. Central and Eastern Europe had other preoccupations, though Warsaw's *Gazeta Wyborcza* had as its second front page story 'Save the Earth, but with what?' – after a call for a movement for defence of the third Polish republic. Russia's *Pravda* and Estonia's *Baltic Independent* didn't seem to mention it at all, among concerns about economic reconstruction and border conflicts.

South Africa's *Cape Times* had only half an inside page, with a front page devoted to local issues, and the Queen Mother; Kenya's *Daily Nation*

Rio 'Newspeak'

THIS BREATHING WORLD – D'vora Ben Shaul

READING through the mountain of material faxed to me from the Rio Environmental Summit, I had the feeling of somehow having wandered into a scene from Orwell's *1984*.

It was all in 'Newspeak'. Good was bad, bad was good, up was down, and so on.

It is amazing to what degree the industrialized nations have managed to convey the impression that, in the end, it is the underdeveloped nations of the world who are completely responsible for excess carbon emissions, global warming and the greenhouse effect, with all the expected climatic changes.

Having spent centuries plundering the natural resources of the planet and polluting its atmosphere, these same industralized nations, choking in their own waste, are – now that the crunch has come – calling for a halt further carbon emission.

Not, of course, at home. No one is really willing to cut back on the amount of carbon produced by fossil fuels, because this would mean cutting back on energy use and industrial processes. The industrialized nations want the undeveloped parts of the world to stop their struggle to obtain a share of the earth's goodies, to be content with what they have and protect their natural resources for the good of all the planet.

These same industralised nations are also calling for population growth control – but here again the target is the Third World, not back home in Europe or in North America.

At the same time, certain things get ignored, like the fact that a child born in the industrialized world will consume many times more units of the world's resources and will produce four to five times as much pollutant as a child born into the undeveloped part of the world.

Some people had great hopes for the congress in Rio, and some of us still have them. But it isn't going to be the jabbering jargon and upside-down morality that brings about any real change in our environmental policy.

▲   *Figure 4.8   Extract from the* Jerusalem Post.

included rather more, with 'Japan pledges aid to developing nations' indicating a public relations success. In the Middle East, Egypt's weekly *Al-Ahram* had a front page story 'US faces growing isolation in Rio', after two Egyptian items. Turkey's *Newspot* had as its first story what Prime Minister Demirel said in his Rio speech. Iran's *Tehran Times* also had a front page news story, after an extremely vitriolic attack on the big powers by the Minister of Culture and Islamic Guidance. Israel's *Jerusalem Post* only had an inside news story, but two features: 'Earth Summit gets down to business matters' about a pre-summit meeting in Rio of the International Chamber of Commerce, and an amusing 'Rio "Newspeak" ' (Figure 4.8).

Returning to Europe, UK views can be compared with treatments elsewhere. Germany's Hamburg weekly *Die Zeit*, Austria's *Wiener Zeitung*, Italy's *Il Corriere della Sera*, France's *Figaro*, Spain's *El País*, and so on, all gave full news and comment that emphasised missed opportunities and disappointment from environmental bodies, as well as some optimism that such a debate about world concerns had happened at all, with little sign of the sourness of some UK and US responses. The Vatican's *L'Osservatore Romano* characteristically gave no account of the controversies at all, but simply quoted in full the speech of its own delegate, including a statement on population policy (Figure 4.9).

Most strikingly, the French monthly *Le Monde Diplomatique* devoted an eight-page supplement to a detailed philosophical analysis of the issues. An introductory essay by the environment minister was followed by closely

When considering the problems of environment and development one must also pay due attention to the complex issue of population. The position of the Holy See regarding procreation is frequently misinterpreted. The Catholic Church does not propose procreation at any cost. She keeps on insisting that the transmission of and the caring for human life must be exercised with an utmost sense of responsibility. She restates her constant position that human life is sacred; that the aim of public policy is to enhance the welfare of families; that it is the right of the spouses to decide on the size of the family and spacing of births, without pressure from governments or organisations. This decision must fully respect the moral order established by God, taking into account the couple's responsibilities toward each other, the children they already have and the society to which they belong. What the Church opposes is the imposition of demographic policies and the promotion of methods for limiting births which are contrary to the objective moral order and to the liberty, dignity and conscience of the human being. At the same time, the Holy See does not consider people as mere numbers, or only on economic terms. It emphatically states its concern: that the poor not be singled out as if, by their very existence, they were the cause, rather than the victims, of the lack of development and of environmental degradation.

Serious as the problem of interrelation among environment, development and population is, it cannot be solved in an over simplistic manner and many of the most alarming predictions have proven false and have been discredited by a number of recent studies. 'People are born not only with mouths that need to be fed, but also with hands that can produce, and minds that can create and innovate.'

▲   *Figure 4.9   Extract from* L'Osservatore Romano.

argued pieces on 'six perils for the planet', 'citizens of the world-city', third world debt, limitations of the market, 'ecologised' thought, planetary solidarity, a new demographic equilibrium, science and economics, NGOs, religion – a serious and thorough treatment contrasting strongly with some of the limited and complacent responses elsewhere.

2.4   Summary

Both the UNCED event itself and world comment on it were of bewildering complexity and diversity. What the comment shows is, understandably, the divergence between rich and poor, but also a whole range of national perceptions and rivalries rooted in culture, history and geographical accident. Perhaps they make the failures of the conference easier to understand and its successes the more to be valued. There was agreement that *some* precautionary action should be taken to reduce the risk of climate change and loss of biodiversity, with continuing monitoring and renegotiation procedures. 'Sustainable development' as an ideal, whatever its uncertainties, was endorsed, though the possible implications for Northern consumption patterns, and for world industrial, trading and financial structures, are hardly beginning to be appreciated. Whether an increasingly globalised culture can overcome the deep-seated inequalities in the world without destructive conflict remains to be seen. At least the Rio conference was a *debate*, however vociferous, not a battlefield.

3   Change and equilibrium

Having gained this broad impression of current attitudes to environment and development, we now turn to our second question:

What processes brought us to this position?

Some of the history of human environmental effects and attitudes needed to answer this question has already been given in the earlier chapters. Here this account is reviewed and extended from a different perspective, which may help us to distinguish those factors dividing nations from those shared in common. An elementary idea, which at least provides a basic framework for analysis, is that all separate events can be located more or less precisely in time and space. The interactions between events, and the social responses to them, may well be very complicated, but it will be a start if we can order events in the right sequence and the right locations. This still covers a large amount of history and geography – so we need a procedure for taking a broad view. This can be provided by recognising a limited number of different *scales*, which each imply different types of process. Try not to confuse these scales, but you do not need to concern yourself with the details used to illustrate them – they are only indicative, and can be extended without limit.

3.1   Time – the ten scales of history

We may be inclined to regard 'history' as mainly concerned with human society, while the physical origins of the Earth are 'geology' and the development of life on Earth is 'biology' and 'evolution'. However, there is something to be gained by considering them all as aspects of one process. We have also so far in this book seen history as moving *forward* in time from a supposed empty world, which we can't readily imagine. The alternative perspective taken here is to look *back* from the present world which we know – strictly speaking, it's all we can ever do; the record of the past is always incomplete, and we can only interpret it from the present. Any assertions that follow are to be understood as the current received wisdom, which is always open to revision in the light of new information or interpretation – this will be particularly true of the longer timescales, where there have been major revisions in recent years. The details chosen for inclusion only represent a personal judgement of importance, and are roughly structured under population, politics, environment and development for the shorter scales, and geology and evolution for the longer ones. Remember that many volumes have been written about each of these scales, and only the sketchiest outline is given here. Distinctive processes, which are still going on, occur on each of these different scales. They provide some sense of the different kinds of things we may expect to happen in the future, some of which are pointed out.

The scale of a human lifetime makes us particularly interested in periods of a hundred years or so, when in fact the further away events are in time, the less likely it is that short periods will matter, at least as far as present consequences are concerned. So the ten scales all reach back from the present by progressively longer periods – for convenience chosen as years, multiplied by ten each time.

1   *The last 10 years (or 10^1 years*) – 'current affairs', a decade of global environmental concern*

*If you are not familiar with this 'power of ten' notation, you will find an explanation in Chapter 5.

	1985	1986	1987	1988	1989	1990	1991	1992	1993	1994	NOW
population:	4842m ------					▶				------ 5630m	
										Cairo conf	
politics:	Reagan, Thatcher, Gorbachev			Bush			Gulf War	Yeltsin, Major, Clinton			
Europe:	EEC		Single European Act			Germany reunited		USSR breakup		Yugoslav conflict	
environment:		Vienna ---------	Montreal ------------			London (ozone)					
		Chernobyl				Geneva (climate) --------------------------------- Berlin					
development:			Brundtland					Rio (UNCED)			

This is a period with which we can all identify, though of course our memories of it will be very different. Population continues to increase, and was the subject of the 1994 UN conference in Cairo. The 'political' developments listed are those familiar from the UK – UK, US and Russian leadership, the end of the Cold War, the continuing evolution of the European Union and painful readjustments in the former Communist countries. However, the topics you would remember if you lived in South America, Africa, China, India or Cambodia would no doubt be very different. There have been bloody conflicts, famines and natural disasters worldwide, some, like the Gulf War, with more international dimensions. Environmental disasters like the toxic chemical release in Bhopal (in 1984) and the nuclear reactor failure in Chernobyl raised both technical and social questions, and showed how effects could range from the local to the regional scale. Furthermore there has been a growing concern about *global* environmental issues. Damaging depletion of the ozone in the upper atmosphere was addressed in a series of agreements, in Vienna, Montreal and London, to reduce and finally abandon production of the chemicals causing it. The more general effect of human activity on global climate was explored in Geneva in 1990, leading to the initial agreement to control relevant emissions at Rio, developed further in 1995 in Berlin. Sustainable development we have already recognised as the issue raised by Brundtland and pursued at Rio. These are all topics to which we shall return in later chapters.

On this scale, day-to-day events are dominated by economic, political and in some cases military conflicts between sovereign states. But global issues are being addressed, as they had begun to be before this decade. How far this debate has yet resulted in significant action can be questioned, but at least all governments and international institutions must include *some* consideration of these global environmental questions in a way they did not thirty years ago.

Activity 4

Draw your own ten-year timescale, on a large sheet of paper, reconsidering the significant events. Keep this by you and add to it as you progress through the book.

(This short period is obviously strongly dependent on when you read this text, which was written in 1995. Extend your scale to the year 2000 to keep new developments in perspective.)

Do the same for the scales which follow. They will be useful for integrating material from different sources.

2   *The last 100 (10^2) years – technological change, population growth, a century of global conflict*

	1900	1910	1920	1930	1940	1950	1960	1970	1980	1990	NOW
population:	*c.* 1600m -----▶										
politics:	European colonial empires - - restructured		WW I	USSR	WW II	dissolved	Cold War		oil crisis	USSR collapse	
									North/South tensions		
environment:	industrial pollution					(UK) Clean Air Acts			UNEP	Chernobyl	
development:	radio	film			TV	nuclear	computer			UNCED	
	car		aircraft			DNA	space				

We all have an image of the events of this turbulent century, from our own memories, those of our parents and vast – if often biased – influence from the developing communication media. The world population has increased by about four times and energy use by about ten times. (The population is becoming stable in industrialised countries, and the optimistic view is that economic development will lead to global population stabilisation over the next hundred years or so.) There have been striking changes in technology, sketched above roughly by the dates particular developments became significant – obviously each of them has a complex history of its own, and you will be able to think of others. The world was already unified, in a sense, at the beginning of the century by trade, steamship, railway, telegraph and imperialism, but the changed availability of transport and telecommunications has enormously increased our awareness of the cultures and problems of other countries, along with the integration of the world financial system ('globalisation'). When the industrial powers came into political conflict, the application of their technical and mass-production methods to warfare and oppression led to the horrific suffering of the two world wars, as well as innumerable smaller conflicts, an expanding international arms trade, and continuing fears of nuclear war. The tensions of the factory system had already led to political opposition to capitalism in the nineteenth century, with communism becoming dominant in Russia after the first world war, and spreading its influence after the second until its recent collapse; other versions of socialism have variously influenced government elsewhere. The highly developed colonial empires of many European countries were restructured after the first war, and dissolved after the second, though their influence remains strong in the financial and trading structures of the present North/South divide, and the consequent differing attitudes to environment and development (within and between countries).

Global issues have been increasingly recognised. For example, maritime trading and economic regulations evolved from the late 1940s through increasing concern with pollution and other environmental questions to a comprehensive Law of the Sea in 1983. Meadows *et al.*'s *Limits to Growth* and Schumacher's *Small is Beautiful* in the early 1970s had questioned the sustainability of the present industrial order. The first UN Conference on the Human Environment in Stockholm 1972 led to the formation of the UN Environment Programme (UNEP), followed by the *World Conservation Strategy* (1980), the Brandt reports on *North–South* (Brandt & Sampson, 1980, and UN Commission, 1983), Brundtland and Rio.

On this scale we see the origins of some of the present international conflicts – competition between capitalist countries; the Cold War between capitalism and communism, and the subsequent instability in Eastern Europe; the continuing legacy of colonialism and the associated trading and financial system.

Apart from our particular perceptions of the present century, this timescale, not much more than a human lifetime, biases our view of earlier times – it will be useful to continue to remind yourself of the number of human lifetimes contained in the longer periods which follow.

3 The last 1000 (10^3) years – 'modern history', global unification

	1000	1100	1200	1300	1400	1500	1600	1700	1800	1000	NOW
population:	300m? - - - - - - - - - - - - - ►										
politics:	Europe, Islam, Mongol China, Africa, America (separate civilisations)					colonisation Slave trade, capitalism		revolutions - - - - - - - - - - - - - questioned		globalisation	
environment:	deforestation					urbanisation		industrial pollution			
development:	agriculture, craft, trade				exploration			steam engine			

Much of our school history, with all its misconceptions and national prejudices, is concerned with this period. (You may recall the humorous *1066 and All That*, which beautifully captured the *English* schoolchild's fractured version of history – it would be salutary to see corresponding accounts from other countries of the world.) During this period the human population seems to have increased by some twenty times. From a global perspective, perhaps the most important transition was from *separate* civilisations in all parts of the world, with only limited trade contact at the margins, to the unified world trading and communication system of the present. From a European point of view, the middle of the period saw the voyages of exploration and 'discovery' – which usually meant discovering other people already there to be traded with, or sadly often enslaved or exterminated. The colonial empires, with their continuing effect on trading patterns, resulted. (Detailed factual information about these developments, and current world inequalities, can be found in Thomas's *Third World Atlas*.) The financial structures of capitalism evolved in parallel over several centuries (Braudel's three-volume work *Civilisation and Capitalism: 15th–18th century* gives a thorough account). Governments have ranged from feudal and authoritarian to democratic and totalitarian, with the emergence of new merchant and industrial classes leading to revolutions of various kinds. Religious constraints on behaviour have tended to be replaced by economic ones, though various forms of religious fundamentalism continue to be influential. Adam Smith in 1776 laid the foundations of classical economics, while the communism of Marx, the socialism of William Morris and the anarchism of Kropotkin in the nineteenth century variously questioned capitalist developments (see reference list).

On this scale, the origins of 'modern' trade and industry, and resulting deeper seated sources of conflict within and between nations, can be traced.

4 The last 10 000 (10^4) years – 'ancient history', the current interglacial period (the Holocene)

	8000BC	7000BC	6000BC	5000BC	4000BC	3000BC	2000BC	1000BC	0	1000AD	NOW
population:	4m? (everywhere) - - - - - - - - - - - ►										
politics:	settlements		Middle East			empires Egypt, China, India, Africa, America			Greece, Rome	Christendom	
environment:	retreating glaciers			deforestation		soil degradation?			pollution		
development:	Stone Age	agriculture, craft, trade				cities pyramids Bronze Age, Iron Age					industry

In this period, the limit of recorded human history, the population seems to have increased by some 1500 times. The period began with the most recent appearance of significant change in the Earth's climate, the end of the last glaciation. The average temperature rose by several degrees over a thousand years or so, and the glaciers covering much of Northern Europe, Asia and America retreated; climatic conditions have since remained relatively constant. This period qualifies for a geological name, the Holocene (though earlier named geological periods are all much longer). Human hunter–gatherers seem to have first established settled agriculture, domesticating plants and animals, in the Middle East at the beginning of this time. Larger settlements led to the complex subsequent history of more elaborate social structures and technology, empires rising and falling worldwide. Removal of forests, intensive agriculture and the growth of cities have significantly modified the natural environment, and possibly contributed to the collapse of empires, as we saw in Chapter 2.

This seems the right place to change from our usual BC–AD terminology, which is culturally determined by the origin of Christianity 2000 years ago, to the more neutral BP (before present). Most of the present major world religions had their origins in the last 3000 years.

On this scale we recognise the transience of local empires, the massive increase in human population supported by agriculture, trade and industry, significant changes in beliefs and value systems, and gradual globalisation of technology and economics.

5   *The last 100 000 (10⁵) years – the last glaciation, human dispersal*

	100 000 BP		50 000 BP			NOW
geology:	upper Pleistocene		glaciation			Holocene
humans:	Africa, Europe, Asia tool-making hunters		Stone Age	---➤ America, Australasia		agriculture

This is mostly the period of the most recent glaciation, with low average temperatures until the last ten thousand years, and further detailed climatic changes influencing the possibilities for human habitation. This is the province of archaeology and anthropology, with complex regional variations. Some of the lifestyles of nomadic hunter–gatherers have persisted in isolated regions to the present, and give some idea of social structures and constraints which provided a stable existence (for a much more limited population) over many tens of thousands of years. Flint tools of growing sophistication were developed; cave paintings provide hints of co-operative hunting rituals as an important part of our pre-agricultural heritage. Their influence on the survival of other large mammals was considerable, so humans were already having an environmental impact.

On this scale we recognise the adaptability of humans to diverse climatic conditions and the importance of co-operation, myth and ritual in human society.

6   The last million (10^6) years – the present Ice Age, the approximate limit of human history

	1 000 000 BP		500 000 BP		NOW
geology:	Pleistocene -		periodic glaciations with shorter interglacial episodes		
humans:	Homo erectus - - - - - - - - ▶		Homo sapiens		
	Africa stone tools, fire?		Europe, Asia		blade tools

During all this time there were fairly regular glaciations, each lasting about 100 000 years, with briefer interglacial periods when the temperature was more like that of the present. The *pattern* of variation seems to correspond to minor fluctuations in the Earth's movement around the Sun, but the *amount* of variation depends on some resulting effect on the Earth's climate system which is not yet fully understood. This pattern seems to have persisted for a rather longer time, about 2 million years, the so-called Quaternary or Pleistocene period, before which conditions on Earth were generally warmer. For this reason, the *whole* period can be described as the *present* Ice Age.

There is no obvious reason why this pattern should not persist, with the present interglacial giving way to another glaciation over the next few thousand years. In this case human society will have to adapt as it has done previously, although in the past with nothing like so large a population or such complex settlements and institutions. Present concern about human-induced global warming over the next *hundred* years has to be seen against this background of 'natural' variation – a crucial question is whether such warming will take us outside this established Ice Age pattern, and what the consequences of that would be.

Study of human remains from around the world leads to varying views about *exactly* when and where 'modern' humans emerged, but there was a related species, *Homo erectus,* using stone tools, and possibly fire, in Africa at the beginning of this time. The colonisation of Europe and Asia, the development of improved blade tools, the controlled use of fire and the transition to *Homo sapiens* occurred during this period.

On this scale we recognise the unity of the human race, with a small population adapting to significant regular climate changes, and the exceptional population growth of the Holocene possibly upsetting that regularity.

These first six scales have given us a view of *human* history which emphasises different kinds of process. The way each is embedded in the next may help in appreciating the even vaster sweep of geological and evolutionary processes, as at present understood. The further four scales required are briefly outlined overleaf; some of the terminology and ideas will be clarified in the next chapter. We see from this that bacterial life, of comparatively simple structure but diverse biochemistry, appears to have been present for about three-quarters of the Earth's history. All other forms of life, more complex, but with a common basic chemistry, have existed for about one-quarter of it.

This then is the barely imaginable (terrestrial) backdrop against which our human adventures are enacted. This is not yet the *cosmic* scale, though a current belief about the origin of the Universe implies that only one further scale is needed to accommodate the 'Big Bang' some 20 000 million years ago. But this raises much deeper questions about the meaning of time itself – if they interest you, they must be studied elsewhere.

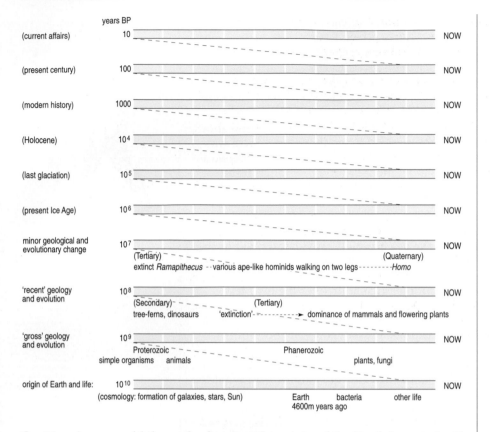

Q   How long would the scale showing the origin of the Earth have to be if each year was shown as on the shortest scale – say, 1 year = 1 cm?

A   10 000 million × 1 cm = 100 000 km, more than twice the distance round the Earth.

Some of you will recognise that these ten – or eleven – separate scales can all be compressed on to a single logarithmic scale ($10^{11} - 10^1$ years), which can also be extended indefinitely to shorter times to cover many details of biological, chemical and physical processes.

The perspective provided by these different *time* scales can be complemented by a similar series of *spatial* scales, from the local to the regional and global. Again, characteristic processes can be recognised on each scale.

3.2   Types of process

Can we use the evidence of change on all the timescales presented above to predict the future, or to guide our policies?

Among the apparently random and unpredictable behaviour, it is useful to identify three types of comparatively *regular* process:

1   Average conditions remain relatively *constant* over some period.
A good example is the average climate of the Earth since the last glaciation. In spite of large variations each day, over the seasons, and over different regions, the average temperature has remained within about 1°C over the last ten thousand years. The language of mechanics provides useful

terminology in the idea of **equilibrium**, of balance between different effects. This can be 'static', when nothing changes at all. This in turn may be 'stable', like a ball at the bottom of a bowl, which will return if disturbed; it may be 'unstable' like a ball balanced on an up-turned bowl, which will roll down if disturbed; or it may be 'neutral', like a ball on a flat surface, which will stay in any position to which it is moved. Or, more relevantly, equilibrium can be 'dynamic' when two or more continuous processes just compensate for each other – like a bath being filled at the same rate as the water is running out: the water is constantly changing, but the level remains constant. The population of a region will be in dynamic equilibrium if the birth rate + the rate of migration into the region is just equal to the death rate + the rate of migration out of it (see Chapter 7).

2   Conditions *recur* in a regular pattern. The most obvious example is the pattern of days and seasons associated with the movement of the Earth around its axis, and around the Sun. More controversial are attempts to recognise longer-scale patterns in history ('trade cycles', cycles of innovation, or of imperial growth and decay, whether or not associated with external effects like the 11-year cycle of sun-spot activity). On a longer scale still, the pattern of recurrent glaciations in the present Ice Age is believed to be linked to systematic long-term variations in the Earth's motion around the Sun.

3   There is a consistent change (increase or decrease) *in one direction*. Here the clearest example is the increase of human population over the last ten thousand years, still continuing. It is useful to distinguish three particular forms of 'growth' (or decay) of this kind: linear, exponential and logistic.

  ● In linear increase the same amount is *added* during each time period.

  ● In exponential increase the quantity is *multiplied* by the same factor in each time period. For example, it might have a constant 'doubling time', of thirty years, say. This is the 'compound interest' process, which leads to a characteristic runaway effect. Human population increase has in fact been 'super-exponential' with the multiplying factor itself increasing with time, or the doubling time getting shorter.

  ● A logistic increase begins exponentially, but then slows down and finally becomes constant as some limit is reached. The normal growth of an organism is of this kind. Population in industrial societies appears to be stabilising in this way, and there is some hope that similar economic and social developments will stabilise the global population over the next century or so.

A policy seeking economic 'growth' does not seem to be compatible with one seeking 'equilibrium'. Of course, there are longer period changes outside human influence to which we must adapt, but the comparatively short 'recent' period of population and economic growth may be disrupting a period of climatic equilibrium.

Activity 5

Consider the changes in the conditions of human life on Earth that might be expected over the next 10, 100, 1000 years. Try to distinguish between essentially unpredictable matters and broader population and climate questions.

3.3   *Humans on Earth: triumph or crisis?*

As a postscript to this discussion of the scales on which we can examine life on Earth, we can remind ourselves of the human role.

Q   For what proportion of this supposed history of the Earth have 'humans' existed?

A   Approximately 1 million years out of 4600 million years – about 0.02%.

Q   For what proportion of human existence have settled agriculture, cities and modern institutions existed?

A   Less than 10 000 years out of 1 million years, i.e. less than 1%.

The development of tool use and social organisation by us, the human animal, led to success as hunter–gatherers for about 99% of our time on Earth. These characteristic human *skills* are thus perhaps a million years old.

They should be distinguished on the one hand from the pre-human *drives* or instincts built into our genetic constitution, many shared with other animals, and evolved over a thousand million years – for survival, reproduction, co-operation, competition, territory, power, aggression, altruism and so on. These are controversial issues which sharply polarise opinion. They will not be pursued here, except to refer anyone interested to Mary Midgley's *Beast and Man: the roots of human nature* (1979). This provides a searching discussion of the relationship between studies of animal behaviour and traditional philosophical debates about nature and nurture, the problem of evil, etc.

On the other hand, our *value* systems have developed, particularly over the last ten thousand years, from earlier family and tribal ones into the diverse religious, political, legal, economic, aesthetic and ethical systems of our varied Holocene cultures. The associated 'recent' technological developments of agriculture, trade and industry have supported what biologists would call a 'bloom' of population growth which evidently can not continue indefinitely. Perhaps the starkest environmental question of all is whether this development, into which we are all inextricably linked, should be seen as a creative triumph – after all, it embraces the pyramids and New York, Shakespeare and television, Buddha and Newton – or as a wrong direction which will be ultimately self-defeating.

In practice, most of us do not adopt either such extreme position, valuing the best of human achievement while seeking to avoid disaster. Our social organisation has developed to provide not only necessities but luxuries for many of us – indeed, much of history has seen the luxuries of one generation perceived as the necessities of the next.

Meanwhile, the 'natural' hazards of famine, flood, earthquakes, volcanic eruptions continue to remind us of the physical constraints to which we are still subject (e.g. Plates 13 and 14). There are obvious geographical differences in these hazards, but they are associated with social differences too. Famine in particular is now more often the result of war or conflicts over land rights rather than unavoidable harvest failure (though the possibility that an increasing population can outstrip the available land, or crop yields, always remains). The rich can better afford flood protection and earthquake-resistant buildings, or to choose where to live.

Not only in these hazards but also in the access to natural resources there is a deep-seated inequality within and between nations which is the mainspring of political conflict and the demand for 'development' to bring to all the benefits enjoyed by some. Human society has increasingly modified the physical conditions on Earth, by agriculture and infrastructure, to improve its chances of survival. It can be argued, as one element of the Gaia hypothesis, that life in general has interacted with environmental conditions, in the atmosphere for example, in ways that have ensured its survival. However, it is increasingly clear that human activities are producing unforeseen changes – to non-human life and perhaps to climate. The question is whether the changes intended to improve human survival chances have been at the expense of stability achieved over longer periods by the global system. This is the deepest reason for 'environmental' concern – not for the survival of life on Earth, but of human society as we understand it, or can imagine it.

3.4   Summary

Ten timescales, each ten times longer than the last, provide a basic framework for locating the physical, evolutionary and human events of terrestrial history and geography. Characteristic processes and patterns can be recognised on each scale. The human history with which we usually identify – that of the last ten thousand years – has produced major technical and social change. Continuing inequalities lead to a demand for further development, but environmental change may be threatening the stability of the global system.

4   Population, resources and technology

Now that we have established a general framework for describing past and present processes, we can begin to consider the question:

What constraints are implied for the future?

4.1   Population

We have already emphasised the increasingly rapid human population growth of the last ten thousand years, only made possible by the technical changes in agriculture, trade and industry over this period. It is generally recognised that this increase cannot continue indefinitely, though opinions differ about the ultimate human 'carrying capacity' of the planet and how far even the present level threatens the stability of the global system. It is believed that appropriate forms of development and social change will lead to global population stabilisation over the next century, as is already happening more locally in industrialised societies. However, this local stabilisation has been associated with a huge disparity in resource use, as we shall see. The perceived threats to the global system, from climate change to resource depletion and waste disposal, arise much more from the high consumption of these stabilising populations than from elsewhere. Whether global stabilisation can be achieved without this increased resource use is a more open question.

4.2   Resources

Like other species, humans make use of the world to provide land and materials which we describe, in our anthropocentric way, as **resources** to ensure survival and comfort. However, unlike other species, humans use a range of resources which is very wide – not only land, plants and animals but also many types of fuels and minerals – oil, tungsten, etc. An elaborate trading system, with its abstract support in an economic and financial system of global scale, has been set up, with its corresponding move from rural to urban societies. Ecologically, a major distinction between humans and other species is that, now, we do not live off our territory – humans are not only upright and naked (or clothed) apes, but can increasingly be seen as distinctively 'trading apes'.

A difficulty about equitable human resource use has been described as 'the tragedy of the commons' – a vivid phrase popularised by Hardin in 1968, though originating in the eighteenth century. The basic idea is that if a resource, like land for grazing animals, or a lake to fish in, is freely available to everyone, it is in each person's interest to continue to use it well beyond the stage where it is being damaged by over-use, since the individual advantage is clear but the collective disadvantage is shared. A modern example would be the use of a private car in congested traffic conditions – the personal advantage may outweigh the small delays experienced individually, even though it is contributing to a large total disadvantage – until the delays become overwhelming.

Of course most resources are *not* freely available to everyone, and self-interest is not the only human characteristic. These factors can be recognised in three kinds of historical response to common resource use:

1   *Exploit and move on*   This is the basis of 'slash and burn' agriculture, or of timber logging and fisheries – if there are enough other places to move to, and enough time is left for the system to recover before returning to it, this is 'sustainable' in the long term, but seems increasingly not to be the case on the global scale.

2   *Privatise*   Most resources have in fact found their way into limited ownership – originally by conquest or theft, more recently by purchase or dexterous legislation. It can be argued that an individual resource owner has an interest in looking after it – many landowners would claim to show 'stewardship' for their land to ensure that it is not over-exploited, though this does not avoid obvious political conflict over how much of any resource should be under one person's control. (Land ownership in particular is a basic source of human class distinctions, revolutions and wars). Once again on the global scale it is not clear that large-scale private ownership, apart from its inequity, has avoided the effects of greed and short-term profit making.

3   *Regulate or tax*   This aims to reduce use to a stable level, and is the 'community' solution. Throughout history local traditions, rules and rituals have often been established in small communities to regulate resource use and control individual greed. National governments continue to make

various attempts to control the use of their 'own' resources by law or taxation. However, the global commons – the sea, the land, the atmosphere – has not been protected in this way, from extraction, misuse or pollution, and the development of *international* control regimes may be our best hope for the future.

4.3   Technology

Technology is a term best used for the totality of all the practical ways a particular society provides for its needs, as in 'Stone Age technology', 'Roman technology' or 'twentieth-century technology'. It is sometimes used as though it were synonymous with 'engineering', the systematic design of useful artefacts, though this is perhaps too narrow. It also includes craft and tradition, the science that provides better understanding of nature and society, and social factors like values, beliefs, industrial structures, profit motives, and the manipulation of opinion in advertising and consumption. It is also used rather weakly, particularly in electronics and computer jargon, for particular ways of doing things: 'amorphous silicon technology', 'MS-DOS technology' and so on, where 'technique' would do as well.

However, the classification of all these newer developments together as 'information technology' does point in an important new direction. Most of our earlier industrial and domestic machinery has used ever more energy, reflected in the ten-fold increase in world energy use in the twentieth century, when population has increased by (only!) four times. More attention is now being paid to the way these services can be obtained with reduced energy use, and with less damaging ways of supplying this energy in the first place. But information technology is also offering a huge increase in access to knowledge and communication, without much expenditure of energy in the equipment itself, and reducing the need for transport. This is already changing our lives and demands on services, and may lead to very different social structures over the next century.

4.4   The PRQ formula

The total environmental impact of the use of a particular resource by all humans on Earth must depend on the product of three factors – the total human population P, the *average* resource use per person R and an index of the *quality* of the technology used to exploit the resource Q – i.e. on $P \times R \times Q$. (This approach is similar to that of Paul and Anne Ehrlich, whose works give more detail.)

It is easier to attach numerical values to P and R than to Q. Thus, for example, in the case of *energy* use, introduced as a distinctive indicator of human social organisation in Chapter 2, the present average rate of energy use per person (R) is about 2 kW (2000 watts).

Energy is not a simple resource. It is a generalisation from separate natural sources like wood, coal, gas and oil, nuclear power, solar energy, etc., with a single measure of quantity (the joule, symbol J) and of *rate* of use (power, measured in watts, symbol W). The 2 kW power rating is the equivalent of every man, woman and child running one large electric fire each at all times (but it applies to all forms of energy use, not just electricity). (For more about energy and energy policy see *Blunden & Reddish, 1996*, Chapter 1.)

Q   If the present world population (*P*) is about 6000 million, what is the
    present total rate of world energy use ($P \times R$)?

A   About 12 million million watts (or 12 terawatts, 12 TW) – comparable to
    12 000 very large power stations (of 1000 MW each).

This provides a useful figure for comparison with the known reserves of
fossil and nuclear fuels (which will last tens to hundreds of years at this
rate, depending on the particular fuel considered), and with the rate at
which energy is supplied continuously by the Sun (which is about 10 000
times greater). But it doesn't yet indicate the environmental impact, where
the further 'quality' factor *Q* is needed. At this point, the *PRQ* formula
ceases to be a matter of simple numbers, but becomes more a description,
and a matter of judgement. In principle, we could decide that *Q* should be a
number between 0 and 1 (*Q* = 0 for no impact, *Q* = 1 for maximum impact),
but assigning this number would obviously be a controversial matter. It
would have to take account in some way of the *different* environmental
effects of extracting and distributing coal, gas, oil, and uranium, of
pollution and waste disposal, the greenhouse effect, the risks of accidents,
proliferation of weapon material, the visual effects of wind turbines or
other renewable energy sources – all topics for further consideration.
Expressing all of these in a single index of technological quality *Q* would
depend strongly on value judgements about different kinds of
environmental impact. Similar considerations apply for each resource used
by humans – land, plants, animals, minerals – leading to similar questions
about the sustainable use of the available reserves and comparative
environmental effect. However, the *PRQ* formulation at least has the merit
of identifying three distinguishable factors and their inter-relationship in
environmental considerations.

Such estimates of *total* human environmental impact, however, conceal
the fundamental issue driving the call for development: the *spread* of
resource use about the average, and of available technology, between rich
and poor within countries, but most of all between different countries.
Thus, returning to the case of energy use, the 2 kW global average embraces
a figure of more than 10 kW per person in the USA, 4–5 kW in Europe and
Japan, about 1 kW in China and not much more than 0.1 kW in some non-
industrialised countries. These inequalities are surely highly inequitable
(and can be shown similarly for many other resources as well as energy).
You may like to consider the distinction made here between 'inequality',
which is in principle measurable, and 'inequity', which is a value
judgement about fairness. Such inequity is the principal reason for
demands from developing countries for the right to similar resource use,
for help in introducing technology with less impact – and for resistance to a
simplistic emphasis on population alone as the source of environmental
stress.

The *PRQ* formula emphasises the inter-relatedness of population,
resource use and technological quality. It is easy to demonstrate that a
minority of the world's population in developed countries is
overwhelmingly responsible for present levels of resource depletion and
the emission of greenhouse gases or other pollutants, and this has helped to
improve the bargaining strength of the developing countries of the South in
recent international debate. However, the difficulties in changing current
practices remain considerable.

To stay with energy use again: while there is no doubt that the
developed countries *could* reduce their energy use per head, and its

environmental impact, by a combination of improved energy efficiency and less environmentally damaging forms of energy supply, it will not be easy to achieve a reduction in the 2 kW global *average* energy use, or in its environmental effects, over the next century in the face of inevitable increases in all those countries currently using less than this average. Furthermore, current ideas about global population stabilisation (say, at 10 000 m or more in 2100 AD – the *P* factor) are based on a belief in the 'demographic transition' from low through growing to high stable populations, which occurred in industrialised countries along with an increase in resource use per head (the *R* factor) (*Sarre & Blunden, 1996*). So a continuing increase in the $P \times R$ product seems inevitable.

The South would argue that the onus is on industrialised societies to show that their $R \times Q$ can be reduced, and that the resulting better technology can be exported, before stressing the *P* of other countries. In fact, all three factors are implicated in both 'environment' and 'development', and must be considered together.

We therefore conclude our introduction to these future options with discussions of environmental damage and development possibilities.

5   Future options

5.1   Environmental change or damage?

We have seen plenty of evidence of environmental change. When do we interpret this as environmental *damage?* One possibility, which is pursued in *Blunden & Reddish (1996)*, is to sharpen up the definition of 'environment' to the narrower one used in systems theory, and this will provide further clarification. But for the moment we can stay with the commonsense view of environment used so far, as the surroundings of human beings which provide for our practical and emotional needs. I would like to suggest four levels of concern about environmental change which lead us to call it damage; they flow into each other somewhat, and are arguably of progressively increasing significance.

The first is *nostalgic*. The world we were used to has changed, and the change alone makes us feel insecure. The pattern of streets in our childhood town, or of fields and hedgerows in our familiar countryside, has been destroyed, and we can no longer find our bearings. It may be that the original pattern was not particularly natural; it may even be that the new one is as defensible. Nevertheless, too rapid or brutal changes offend sensibilities, and those who make them at least might be expected to consult those affected, and give reasons why the changes should be seen as improvements. *The World We Have Lost* was Laslett's 1965 comparison of English society before and after the Industrial Revolution. A reminder that this world had major *dis*advantages may help us to accept *some* changes more willingly than others. Nevertheless, disaffection about rapid change of any kind will always remain.

The second is *aesthetic*. The new environment is not only different, it is worse; what was beautiful is now ugly. Such judgements are notoriously subjective – the steam trains that offended in the early nineteenth century are now part of a treasured heritage, as are the mediaeval tithe barns, or country houses sometimes built on the proceeds of the Slave Trade that were originally symbols of exploitation and privilege; even the ruins of old factories may have romantic grandeur. Nevertheless, such arguments about changing standards of beauty can hardly wipe out the aesthetic outrage most people felt about polluted industrial towns before the Clean Air Acts or still feel about the same problem in some Eastern European cities today, about the urban squalor of neglected inner cities, about 'factory farms' or the destruction of treasured landscapes. This outrage may eventually be translated into systematic statistics about health risks, cruelty or social deprivation, but an aesthetic response is an instant recognition of our instincts for what is good for us, and of our social conditioning: a lightning synthesis of many impressions that precedes rational analysis.

The third concerns *human welfare*. The new environment is 'damaged' because it has toxins in the air, water, land or ecosystem that threaten human health; the land is being eroded, desertified or salinated so that it can no longer grow crops; living conditions are such as to cause social unrest. Such judgements require detailed *understanding* of the workings of the physical system, of the flow of materials through it, and of human physiology, psychology and society if actions to correct the damage are to be well judged – though at the same time action may well be necessary while uncertainties remain.

The fourth concerns *global planetary welfare*. Are the changes we are making to improve human conditions in fact threatening the systems, evolved long before human influence, which make the Earth a favourable environment for life of all kinds, and human life in particular? Deforestation, loss of biodiversity, the ozone hole, greenhouse gas emissions are perhaps in this category, and may be the most serious form of environmental damage of all. The understanding mentioned above must now extend to the whole global system, and precautionary action while uncertainties remain may be even more necessary.

Such global scale damage has particularly occurred over the last century, and will require our best interdisciplinary skills to tackle it – from natural and social science, enlightened technology and policy-making on the international and local scale to changes in individual attitudes and expectations.

5.2   Development – growth, sustainability, equilibrium?

Human beings are endlessly active and inventive; this has modified the conditions of life in ways we are all familiar with, and provides the possibility of further development to improve human welfare. But how are we to ensure that in solving one set of problems we do not create new ones?

First of all, we should not identify development with 'growth' – growth was once characterised as 'more stage coaches', while development is 'replacing them with cars'. We should now rephrase that – 'growth' means more cars and more fuel consumption, 'development' means improving the transport system with less fuel consumption. However, 'economic growth' is a policy objective of all governments in capitalist countries – or indeed of many people everywhere – as the source of increasing prosperity. A new school of environmental economists (Pearce, Daly) is trying to reconcile environmental concerns with economic theory. They have pointed out that recognised economic indicators like Gross National Product do not take into account the loss of environmental capital such as natural resources and

amenity, or that activities creating pollution *and* those required to control it
are both included in estimating its value – but such indicators continue to
be used.

The attempt to show in *Limits to Growth* (Meadows *et al.*, 1972) that
existing attitudes were in danger of leading to economic collapse, from
resource exhaustion, pollution and/or population growth, was much
criticised. No doubt the computer modelling used was oversimple, too
insensitive to human ingenuity and the way the system responds to
shortages, too sensitive to the particular initial conditions, and so on.
However, its authors returned fairly unrepentant in 1992 with *Beyond the
Limits*. The essential mathematical point hardly needs computer modelling.
Compound interest processes arise where the increase of something, like a
population, or industrial investment, is proportional to the existing amount,
of population, or industrial capital. This leads to runaway exponential
growth, which is bound to hit limits in a finite world. This idea was first
pointed out by Malthus, in relation to population and food supplies, at the
end of the eighteenth century. In the short term the difficulties were
overcome by colonial expansion and technological change. They are now
recurring on a global scale. The question is always whether stabilising
processes to prevent this runaway are present, or can be introduced in time.

To some extent ideas like theirs have influenced the thinking about
sustainable development of Brundtland and Rio, but it remains to be seen
whether the economic and financial institutions which are so firmly locked
into the 'growth' ideal can change direction sufficiently. For example, an
important requirement of 'sustainability' is that the needs of future
generations should not be jeopardised by our actions now – but this is
immediately threatened by the normal economic planning practice of
'discounting' future costs. According to this, costs sufficiently far in the
future (how far depending on the discount rate used) could effectively be
met by a minuscule investment now. In practice, this is often not done, so
that remote future costs are effectively ignored. Even if, in our economic
system, investments can produce an expected return, which determines the
discount rate, there is still no way this return can be guaranteed. There is
also a deeper question about where this return comes from in the first place
– it may involve resource depletion or environmental degradation.

The strength of the 'sustainable development' ideal is that it does try to
reconcile the real aspirations of the poor to better conditions, which it
would be offensive to deny, with concerns about environmental damage.
A weakness is, however, the anthropocentric view that 'development'
represents 'progress', which can easily be identified with 'growth'. Another
formulation of an ideal future might be to aim for a stable and *equitable
equilibrium* in human demands on planetary resources, which the
developments of Holocene society, and particularly of the industrial society
of the last three hundred years, in no way approach at present. How our
actions can be chosen and our planning procedures developed so as to move
towards such an equilibrium, which is equitable between different people
now, and between present and future generations, will continue to exercise
us throughout this series of books (and no doubt far beyond them).

6   *Summary and conclusions*

We have now discussed the three questions raised at the beginning of this chapter, and can review the answers we have found.

> How are environment and development issues perceived globally in the 1990s?

There is great diversity worldwide, both in general and in responses to the Rio Earth Summit. There is considerable complacency and lack of interest in many industrial countries, though a more enlightened response recognises a critical situation. There is more serious consideration of potential environmental damage in the North than of ways to help redress the development inequalities of more immediate concern to the South.

> How did we reach this position?

A historical survey on various timescales indicates the divisive factors between nations over the last few hundred years, but emphasises the common human problems seen when sufficiently long timescales are considered, and also the global unification processes at work at present.

> What constraints are implied for the future?

There is both a global environment problem resulting from total human resource use, and an equitable development problem resulting from gross inequalities in access to resources.

Two frameworks are established in this chapter: time and space scales and the *PRQ* formula. They lead to a general discussion of environmental damage, development, growth and equilibrium and will help in interpreting the subsequent chapters and books in this series. The remainder of this book takes up the long-term view with a natural science account of the physical system of the Earth, and the life it supports.

Comments on Activities

Activity 1

Your present personal view might be anything between an optimism which sees no problem with population, believing that the extra hands and brains making it easier for food and other goods to be produced, and a pessimism which regards population growth as the principal problem in the world. No doubt the discussions later in this chapter and book will influence this view. If you lived in China you might be aware of being part of the largest national population, a fifth of the world total, and of draconian government attempts to limit family size to reduce population growth. As a subsistence farmer you might be less concerned about global issues than about your personal family size, as a balance between your land's ability to provide food and the extra help, now and in old age, your children can provide. As a member of the Institute of Directors you might be interested in expanding markets, but also in competition from areas of cheap labour, so perhaps in strict international policies to try to contain population growth elsewhere. As a member of Greenpeace you would perhaps consider any emphasis on population growth in developing countries as an unacceptable distraction from the much greater environmental effect of resource consumption in industrial countries.

Activity 2

You may have identified three kinds of possibility, depending on your background: one kind technological, the others more concerned with social policy and organisation:

1   The fuel efficiency of vehicles might be improved; oil could be replaced by liquid fuels derived from coal (not such a limited resource, but still polluting) or plants (a 'renewable' resource); electric vehicles would eliminate local pollution, but the primary fuel source for producing the electricity has to be considered in its polluting effects; hydrogen fuel cells might replace batteries, etc.

2   A significant shift to other modes of transport – cycle, bus, rail, etc. (with their varying forms of fuel use) – might be encouraged by taxation, subsidy or legislation.

3   Employment, shopping, housing, and land use might be planned to reduce the need for transport of all kinds.

The implications would vary according to the choice made:

●   internationally, on the influence of oil-producing countries and international oil companies;

●   nationally, on investment and employment in fuel, vehicle and infrastructure industries (roads, petrol stations, etc.);

●   locally, on the changed environmental effects of the new vehicles and infrastructure;

●   personally, on the changed costs, amenity, mobility offered by the alternatives.

Activity 3

A personal interpretation of the various quotations follows – yours may be different.

The *Sun* leader was a blatant appeal to popular prejudices against politicians, bureaucrats and foreigners, springing primarily from a cynical knowledge that this sells newspapers, but also conceivably as a manipulative device used by powerful interests to discredit anything that might reduce that power.

The Times was evidently more measured and civilised, but still betrayed some of the same complacent superiority: the 'action plan ragbag', 'rapid population growth . . . in some of the poorest countries', 'discredited Third World *canard*', 'the West . . . helping the poor avoid the mistakes it once made', 'did little . . . by consistently pillorying the United States'. It did accept the need for something to be done, but still from a patronising position of Western power, where we have apparently already learned the necessary lessons, and know exactly how to tell *other* people what to do. In this it may reflect a widespread view within the UK, but certainly not that of more modest opinion – that we need to change too, arguably quite radically.

The *Wall Street Journal* is self-evidently the voice of US financial interests, who clearly calculated that Bush's tough stance, if not as prudently packaged as they would have liked, had successfully avoided anything that would require any significant change, or reduction in their influence. This is surely a strong and intransigent strand in US opinion, which, however, includes strong environmental groups of opposing views as well.

The *Japan Times* reflected national perceptions of world leadership, particularly financial and technological, though also apparently in North-South diplomacy and in resource-conscious consumption too. How far this is only the voice of expansive commercial interests, or also embraces national traditions of elegant economy in design, is difficult to judge. The consumerism of Tokyo, the social stresses of industrialisation, the dependence on massive imports (e.g. of tropical hardwoods) and so on are surely not so different from those of the West. But the wish to take Rio seriously, and the optimistic tone, is in striking contrast to the previous quotations.

As was also that of the *Bangkok Post*, perhaps surprisingly critical of 'blanket aid' as opposed to 'self-help and natural development'. The conviction that 'the real issue was always sustainable economic *growth*' betrayed a standard capitalist misreading of sustainable *development* however, and indicated how Southern establishment objectives of following Northern industrial models remain in place. How far these views would be those of Bangkok's poor, or indeed of Thailand's cultural traditions, can be questioned.

Activity 5

One answer – but a very speculative one – would be as follows.

Obviously major catastrophes like all-out nuclear war, or collisions with large extra-terrestrial bodies, are essentially unpredictable. But, within foreseeable futures, the next ten years will surely see continuing global population increase, national economic, political and (it is to be hoped, contained) armed conflicts within and between North and South, periodic natural disasters (earthquakes, floods, etc.), but some progress towards better understanding of climate change and international agreement to try to limit it, along with other global issues (nuclear proliferation, deforestation, resource utilisation and waste disposal, species loss, etc.).

The next 100 years may see either catastrophic climate change or better understanding of why it hasn't happened – or something in between, like the partial success of policies to reduce global emissions. The hoped-for link between economic development and population stabilisation will have been demonstrated or disproved. There will be significant technological and social change affecting resource depletion and waste disposal. World government, either authoritarian or more benignly decentralised, will perhaps be closer.

The next 1000 years is surely beyond political prediction, with the science fiction extremes of complete social collapse, rigidly regimented vast populations, or peaceful small communities quietly co-operating (on the super-Internet?) all possible. However, it will be clearer whether the Ice Age glaciation pattern is persisting, so that society must prepare to adapt to it, or whether it has been disrupted by human-induced climate change.

References and further reading

BLACKMORE, R. & REDDISH, A. (eds) (1996) *Global Environmental Issues*, Hodder & Stoughton/The Open University, London (Book Four).

BLUNDEN, J. & REDDISH, A. (eds) (1996) *Energy, Resources and Environment*, Hodder & Stoughton/The Open University, London (Book Three).

BRANDT, W. & SAMPSON, A. (eds) (1980) *North–South: a programme for survival* (The Brandt Report), Cambridge, Mass., MIT Press.

BRAUDEL, F. (1979) (English translation) *Civilisation and Capitalism: 15th–18th century*, vol. I (1981) *The Structures of Everyday Life*, vol. II (1982) *The Wheels Of Commerce*, vol. III (1984) *The Perspective of the World*, Collins, London.

BROWN, L. R. (1981) *Building a Sustainable Society*, (Worldwatch Institute), W.W.Norton, New York.

BROWN, L. R., Annual *State of the World* reports, Worldwatch Institute.

DALY, H. & COBB, J. (1989) *For the Common Good*, Beacon Press, Boston.

EHRLICH, P. R. & A. H. (1991) *Healing The Planet*, Addison-Wesley, Reading.

GRUBB, M. *et al.* (1993) *The Earth Summit Agreements: a guide and assessment*, Royal Institute of International Affairs, Earthscan Publications, London.

HARDIN, G. (1968) The tragedy of the commons, *Science*, **162**, pp. 1243–8.

LOVELOCK, J. E. (1979) *Gaia: a new look at life on Earth*, Oxford University Press, Oxford.

LOVELOCK, J. E. (1988) *The Ages of Gaia*, Oxford University Press, Oxford.

KROPOTKIN, P. (1899) *Fields, Factories and Workshops Tomorrow*; modern edition 1962, Allen & Unwin, London.

LASLETT, P. (1965) *The World We Have Lost*, Methuen, London.

MARX, K. (1970) *Selected Works of Marx and Engels*, Laurence and Wishart, London.

MEADOWS, D. H. *et al.* (1972) *The Limits to Growth*, (Club of Rome), Earth Island, London.

MEADOWS, D. H. *et al.* (1992) *Beyond the Limits*, Earthscan, London.

MIDGLEY, M. (1979) *Beast and Man: the roots of human nature*, Methuen, London.

MORRIS, W. (modern edition, 1962) *Selected Writings and Design*, Penguin, London.

PEARCE, D. *et al.* (1989) *Blueprint for a Green Economy*, Earthscan, London.

REDCLIFT, M. (1987) *Sustainable development: exploring the contradictions*, Methuen, London.

REDCLIFT, M. & BENTON, T. (eds) (1994) *Social Theory and the Global Environment* (Global Environmental Change Programme), Routledge, London.

SARRE, P. & BLUNDEN, J. (eds) (1996) *Environment, Population and Development*, Hodder & Stoughton/The Open University, London (Book Two).

SMITH, A. (1776) *The Wealth of Nations*; modern edition 1976, Oxford University Press, Oxford.

SCHUMACHER, E. F (1973) *Small is Beautiful: a study of economics as if people mattered*, Blond & Briggs, London.

The Times Atlas of World History (1978), Times Books, London.

THOMAS, A. (1994) *Third World Atlas*, Open University Press, Milton Keynes.

UN COMMISSION (1983) *Common Crisis, North–South: co-operation for world recovery*, (The Brandt Commission), Cambridge, Mass., MIT Press.

UN WORLD COMMISSION ON ENVIRONMENT AND DEVELOPMENT (1987) *Our Common Future* (The Brundtland Report), Oxford University Press, Oxford.

Chapter 5   Earth as an environment for life

1   Introduction

This is the first of three chapters giving a 'natural science' view of life on Earth. It indicates the broad context in which particular environmental processes and problems occur. This context has two aspects. First, the very long history of the Earth, outlined in Chapter 4, which has produced the present physical structure of continents, rocks, oceans and atmosphere, and which can be reconstructed from the geological record. Second, the major circulations of atmosphere and oceans, which combine to create the global distribution of climatic zones. The interaction between climate and geology to produce soils, vegetation and animal communities has already been outlined for Cumbria in Chapter 1 and is explained in greater detail in Chapter 6.

Three major points will emerge from the discussion of physical structure and global processes:

1   A major distinction exists between materials and energy. The materials from which the Earth, ocean and atmosphere are made are, with the exception of incoming meteorites and outgoing space probes, essentially finite. They can be transformed, physically and/or chemically, but not added to. Energy, on the contrary, is dominated by the flow of incoming solar energy and the mechanisms by which the Earth loses energy to space. A minor exception to this distinction is that nuclear technology allows some transformation of matter to energy, but the amounts involved are small in comparison with those involved in natural processes.

2   Although the physical processes in the continents, oceans and atmosphere are major factors in generating the environments for living things, it will be shown that living things have played crucial roles in influencing these physical processes over thousands of millions of years. In particular, the composition of the atmosphere has been almost totally transformed.

3   The major structures and processes which provide the physical environment for life have evolved over geological time, and they still remain subject to change. This is most apparent in the case of climate, but geological and oceanic processes can also bring about major change. Such physical changes would have major implications for living things.

The chapter will start by looking at the history of the solid Earth as a way of showing its enduring basis for life and its very long history, including the role of living things in transforming rocks and atmosphere. The next step is to analyse the contemporary atmosphere, both in its structure and its response to incoming solar radiation. Atmospheric circulation then leads on to ocean circulation. Finally, the chapter considers the cycling of water and carbon between land, air and ocean to demonstrate the strong interconnections that exist between them and between physical processes and living things.

As you read this chapter, look out for answers to the following key questions.
● How has the physical structure of the Earth developed, and what has been the role of living things in the process?
● How does the atmosphere work, and in particular how does it react to incoming solar radiation?
● How do the water and carbon cycles link atmosphere, oceans and solid Earth?

2   The physical structure of planet Earth

2.1   Introduction

The Earth seen from space is a blue planet, with 71% of its surface covered in water. Indeed, it might be more apposite to call our planet 'Ocean' than 'Earth' (see Plate 3).

The Earth of today was born 4600 million years ago by the accretion of gas and dust. In the conditions shortly after this formation, neither oceans nor an atmosphere existed. The Earth's early atmosphere was formed later and was devoid of **oxygen**, being composed of gases poured out by volcanoes, the chief of which (98%) was carbon dioxide. Water made an appearance as steam, also gushing from volcanic vents, and eventually formed oceans.

There is still much speculation about what happened to the Earth after the initial accretion, and about the deep structure of the Earth far beneath the crust which formed on its surface. The zone immediately beneath the Earth's crust is called the mantle (Figure 5.1).

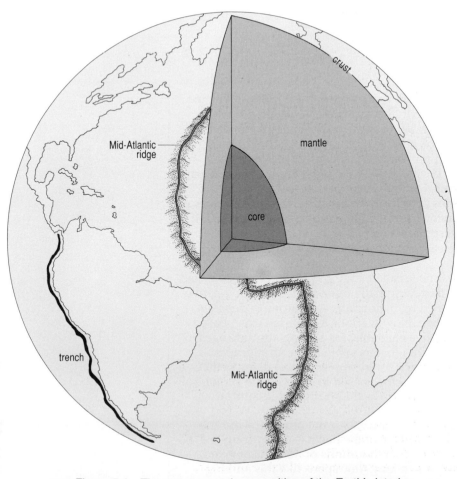

▲   Figure 5.1   The structure and composition of the Earth's interior.

Geologists classify rocks into three general classes, depending upon how they were formed. **Igneous rocks** such as granite are formed from the molten state; **sedimentary rocks** such as limestone and sandstone are formed by the deposition of material from air, ice, or water; and **metamorphic rocks** are rocks of any origin which have been transformed (metamorphosed) by heat which has recrystallised them inside the Earth without melting them – marble and slate are examples.

Q   Of what kind of rock must the crust have been made when it first formed?

A   Igneous rocks formed when the Earth's surface cooled. Igneous rocks still form the bulk of the Earth's crust.

As the Earth cooled, it is reasonable to suppose that water vapour condensed and dissolved gases from the primitive volcanic atmosphere, forming a highly corrosive acid rain that poured upon the newly formed crust. The chemical action of rain on the first igneous rock caused it to weather into fragments which produced the first sediments.

Q   Where would you expect these sediments to end up?

A   In the oceans, where they would be carried by rivers. Most sedimentary rocks were laid down in the oceans, or in river deltas.

2.2   Geological time

The processes which form sedimentary rocks give us an important key to understanding the history of the Earth. Because they are formed by deposition onto existing rock surfaces, younger sedimentary rocks lie in layers over older rocks. If you look at the exposed face of a sand quarry or the rocks exposed by a road cutting you will frequently see a series of layers, or strata. Generally, the deeper you go the older are the rocks. Each rock stratum has its own chemical and physical characteristics, and may bear characteristic fossil remains of animals and plants. All these features of a rock are related to the kind of environment in which it was deposited. If we know how to read and interpret this record in the arrangement of geological strata, then we can learn something about past rock-forming environments.

The arrangement of rock strata is called stratigraphy and has been used to establish a geological time-chart. Originally this only told us the relative ages of different rocks, determined by their relative positions in deposits where they had not been disturbed. Now, various techniques exist, based upon the rate of decay of radioactive elements, for putting absolute ages to igneous rocks, so that the position of igneous strata in the stratigraphic record can be used to attach absolute (though approximate) dates to the geographical record. Because stratigraphy is used to compile the geological time chart, it is conventionally referred to as the stratigraphic column (Figure 5.2).

Don't worry about the details of Figure 5.2 for the moment – but the figure should remind you that the history of the Earth is a very long one! So long in fact that we have to measure it in thousands of millions of years. There is a useful way of writing down long time spans like this and big numbers in general. To see how this is done, look at Box 5.1 now.

◄ Plate 1
Wastwater with Great Gable and Scafell. This view so much typifies the wild face of the Lake District that it has been used in abstract form as the National Park's logo.

◄ Plate 2
Blea Tarn Farm.
The juxtaposition of stone farmhouses, slate roofs, dry stone walls, trim green pastures and rugged fells epitomises the appeal of the Lake District.

◄ Plate 3
Planet Earth as seen by a
European Space Agency
weather satellite (colour
enhanced by computer
processing techniques).

▲ Plate 4   Tundra (Norway).

◄ Plate 5
Boreal coniferous forest
(Labrador, Canada).

◄ Plate 6
Temperate deciduous forest
(Vallets Wood, Forest of
Dean, England).

◀ *Plate 7*
Steppe/prairie grassland
(Colorado, USA).

◀ *Plate 8*
Desert (Sonoran Desert,
Arizona, USA).

▲ Plate 10   Tropical rainforest (Malaysia).

◄ Plate 11
Blanket bog at Moor House
National Nature Reserve,
North Yorkshire.

▲ Plate 12   'The Four Seasons: Summer' (detail) by Nicolas Poussin (1594–1665).

◄ Plate 13
'An eruption of Vesuvius'
by Joseph Wright of Derby
(1734–97).

◄ Plate 14
Volcanic eruption in Hawaii.

▲ Plate 15 'Ullswater from Gowbarrow Park' by J. C. Ibbetson (1759–1817).

▲ Plate 16 'Lake Keswick' by J. M. W. Turner (1775–1851).

Box 5.1   Describing big numbers – powers of ten

The age of the Earth, the amount of water in the oceans, the number of human beings now living on this planet, the distance of the Earth from the sun – all of these important quantities involve measurements that are so large that they are beyond our everyday experience. Nevertheless, we need a way of writing down such numbers that does not involve using a string of noughts, or confusing terms like 'a billion' which means a quantity one thousand times larger in British usage (a million million), than when American writers use the same word (a thousand million). To confuse matters still further, many, but not all, British writers are gradually adopting the American meaning of billion.

The trick for describing big numbers is to convert them to a smaller equivalent number multiplied by a 'power of ten'. A power is the number of times you multiply something by itself. So for example, the number three to the power two is two threes multiplied together: 3×3; the number ten to the power two is two tens multiplied together: 10×10; the number ten to the power three is three tens multiplied together: $10\times10\times10$.

Q   What numbers are represented by the expressions 'ten to the power two' and 'ten to the power three'?

A   Ten to the power two is 10×10, which equals 100. Ten to the power three is $10\times10\times10$ which equals 1000.

Writing 'ten to the power two' is a long-winded expression, so this is abbreviated by writing the power as a superscript number after the ten, thus: 10^2. Another name for this number is of course 'ten squared'.

Q   What number is represented by the quantity 'ten cubed'?

A   Ten cubed is 10^3, which is $10\times10\times10 = 1000$.

Powers higher than three don't have such convenient names as 'squared' and 'cubed', but remember that all powers are calculated in the same way.

Q   Now try converting the following numbers into powers of ten:

10 000;   100 000;   1 000 000

A $10\,000 = 10^4$; $100\,000 = 10^5$; $1\,000\,000 = 10^6$

How did you work out the answers to this question? The quickest way is simply to count the number of noughts that follow the one. If this wasn't the method you used, go back to the question and check for yourself that this method works.

It is very useful to remember that one million (1 000 000) expressed as a power of ten is 10^6.

Although we don't bother writing small numbers in the form of powers of ten, it is worth knowing that even the numbers 1 and 10 can be written respectively as 10^0 and 10^1 (0 noughts, 1 nought).

Now, how do we write large numbers that aren't so conveniently rounded to the nearest whole power of ten? For example, one estimate of size of the human population of China in 1988 was 1 150 000 000. This can be expressed using powers of ten as: 1.15×10^9. Note that we have simply placed a decimal point after the first digit and multiplied this by ten raised to a power. The value of this power is 9, because there were 9 digits following the position where we placed the decimal point. We could also have written 115×10^7.

If you cannot see that 115×10^7 and 1.15×10^9 are the same, note that 10^9 is the same as $10\times10\times10^7$, so you can write 1.15×10^9 as $1.15\times10\times10\times10^7$: this is equal to 115×10^7.
Now try these questions:

Q   A metric tonne is the name given to a mass of 1000 kilograms (kg). An adult blue whale could weigh 177 tonnes. What is this expressed in kg using a power of ten?

A 1.77×10^5 kg

Q   Skiddaw in the Lake District is a mountain 3053 feet high. It is nowadays the scientific convention to express measurements of height and length in metres (m). 1 metre =3.281 ft. What is the height of Skiddaw expressed in metres using a power of ten?

A 9.31×10^2 m

Q   For much smaller measurements than the height of mountains we use millimetres (mm). There are 1000 mm to a metre. What is the height of Skiddaw expressed in mm using a power of ten?

A 9.31×10^5 mm

Now return to the main text.

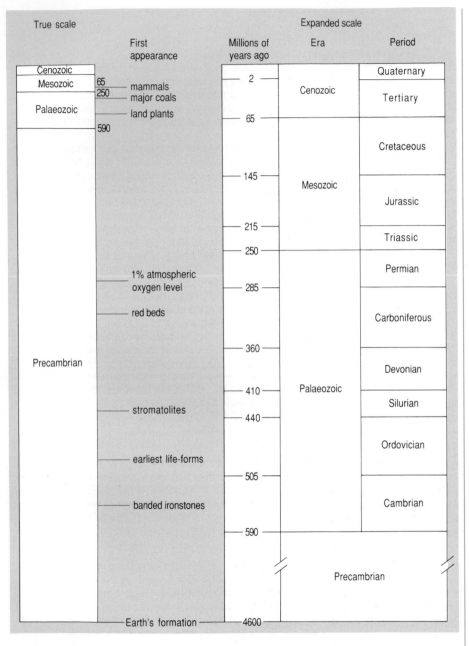

▲ Figure 5.2   The stratigraphic column.

Examine Figure 5.2 now.

Q   What is the estimated age of the Earth? Write down your answer using powers of ten.

A   4600 million years = 4.6×10^9 years.

Q   How long ago did the first life appear on Earth? Write down your answer using powers of ten.

A   About 3.8×10^9 years ago.

Q   What is the name given to the oldest section of the stratigraphic column?

A   The Precambrian.

2.3   *The effects of living things*

The first signs of life appear in rocks dating back to the Precambrian, and these give us some clues about the kind of environment which existed at that time. Curiously enough, another legacy of early life is to be found in the modern environment – in the Earth's **atmosphere**.

The atmosphere of the young Earth in the early Precambrian was relatively rich in carbon dioxide and very poor in oxygen compared with its composition of today (Table 5.1)

Table 5.1   The composition of the Earth's atmosphere today, and as it is thought to have been in the early state, before life evolved

Gas	Present-day Earth/%	Early Earth/%
oxygen	21	0
carbon dioxide	0.03	98
nitrogen	79	1.9

The oxygen now in our atmosphere was then locked up in rock minerals (the compounds of which rocks are largely composed) and in water. We shall see later what happened to the carbon dioxide in the early atmosphere, but first we must take a look at oxygen to understand the significance of this vital element in the chemistry of life.

Oxygen, in its free (uncombined) form as a gas or a solution in water, combines readily with many elements and chemical compounds and changes them. An example is rust, which is caused by the chemical combination of iron and oxygen. Reaction with oxygen usually involves the production of heat. In fact the burning of wood, coal, oil and other fuels all involve the reaction of the carbon compounds in these materials with oxygen, generating heat, water and carbon dioxide. The importance of oxygen to life is that living organisms also 'burn up' carbon compounds in a chemical equivalent of combustion that is called **respiration**. Respiration also produces carbon dioxide and water, but the energy liberated in the process is captured in chemical form rather than as heat and is stored for later use. If the food living things burn to obtain energy was liberated directly as heat, they would have to work like steam engines to make use of it! Ultimately though, this chemically stored energy *is* liberated as heat, as you can discover when you use a lot of it quickly, such as when running for a bus.

The chemical changes known to occur in the presence of free oxygen can be used to interpret the chemical composition of rocks and to determine the atmospheric conditions which prevailed when they were formed.

Not surprisingly, most exposed early Precambrian rocks are igneous or metamorphic. Some of the oldest sedimentary rocks include banded ironstones (see Figure 5.2) formed by the precipitation of certain types of iron compounds from water in the oxygenless atmosphere of the early Precambrian. Banded ironstones disappear from deposits formed in the later Precambrian at about the same time that other sedimentary rocks known as 'red beds' make an appearance. Red beds are so called because of their colour, caused by the presence of rust-like compounds. The disappearance of banded ironstones and the appearance of red beds both indicate an atmosphere that was becoming enriched in oxygen. Where did the atmospheric free oxygen come from?

The answer lies in the first appearance, almost 3×10^9 years ago, of organisms which have left fossilised remains called stromatolites. These are formed from layer upon layer of mats of blue-green **bacteria** which were probably the first organisms to generate free oxygen as a by-product of the process known as **photosynthesis**. Chemically, photosynthesis reverses the process of respiration.

Q   Respiration burns carbon compounds (consuming oxygen in the process) to produce energy, water and carbon dioxide. If photosynthesis achieves the chemical reverse of respiration, what raw materials would you expect it to require as inputs, and what would you expect the products of photosynthesis to be?

A   Photosynthesis requires the input of energy (as sunlight), carbon dioxide and water. The main outputs are carbon compounds and gaseous oxygen.

Today stromatolites are still produced by blue-green bacteria living in the shallow coastal waters of western Australia, the Bahamas and elsewhere. Unlike most other living organisms, blue-green bacteria are able to respire (and therefore unlock chemical energy) *without* oxygen. This, combined with their oxygen-liberating ability, was therefore a crucial step in the creation of conditions favourable to the evolution of organisms for whom oxygen *is* essential.

Apart from the first appearance of life itself, the evolution of photosynthesis in plants and some bacteria such as the blue-greens was arguably the most important event in the whole history of the Earth.

Q   How could the appearance of photosynthesis during evolution explain the change in the composition of the Earth's atmosphere, shown in Table 5.1?

A   Plants use the energy of the sun to turn gaseous carbon dioxide into carbon compounds. In the process, they liberate free, gaseous oxygen. Table 5.1 shows that the modern concentration of carbon dioxide is enormously less, and oxygen concentration considerably greater, than was present in the primitive atmosphere.

Starting with the appearance of stromatolites, photosynthesis appears to have been responsible for the change in atmospheric oxygen concentration through geological time. Plants must also have contributed to the reduction in carbon dioxide concentration in the atmosphere, though the role of photosynthesis in this change was probably a relatively minor one, as we will see shortly.

Because gaseous carbon dioxide is converted into carbon compounds and plant tissues by photosynthesis, the process is often referred to as **carbon fixation**. Once fixed into plants in the form of compounds, the carbon and energy in them can be used to build new plant tissue, or in other words to grow. In the process of growth some of the simple carbon compounds are used to build more complicated ones such as cellulose, which is the main structural material of wood. Growth, and indeed the everyday maintenance of living tissues, requires energy, and this comes from the combustion of carbon compounds.

The atmospheric oxygen liberated by photosynthesis is used in respiration by plants and animals. Plants and animals obtain their energy by respiration, but the food that fuels respiration comes from a different source in each case. Because plants use sunlight and do not rely on other organisms for their source of energy they are called **autotrophes**, or literally 'self-feeders'. They manufacture their own carbon compounds from water and carbon dioxide, using the energy of sunlight.

Animals cannot capture sunlight or make carbon compounds from atmospheric carbon dioxide for themselves, and all their food must therefore come, directly or indirectly, from autotrophes. Animals are described as **heterotrophes** because they feed on others.

Q   Fungi and most kinds of bacteria are unable to photosynthesise. Are they autotrophes or heterotrophes?

A   They are heterotrophes. Many of them feed on dead plant and animal remains.

Q   Carnivores don't eat plants. Where does their food *ultimately* come from?

A   Foxes and other carnivores eat other heterotrophes, but this food ultimately comes from plants which have been eaten by the carnivores' prey.

We will explore the fascinating ramifications of these relationships between living organisms in Chapters 6 and 7, which deal with ecology. For the moment we simply ask: If respiration by animals and plants is the reverse of photosynthesis by plants, how is it that the emergence of plants produced a change in the composition of the atmosphere? Why wasn't the oxygen they produced all turned into carbon dioxide and water by respiration? Once again, the answer lies in the rocks.

Life first evolved in the sea, and so the plants responsible for the rise in atmospheric oxygen must have been marine: they were the **phytoplankton**. These microscopic, free-floating plants exist in vast numbers in all the oceans. They are small but they are able to fix large quantities of carbon dioxide and liberate large quantities of oxygen because they are so numerous and capable of multiplying very fast (see also Chapter 7).

The oxygen liberated by the photosynthesis of phytoplankton and other plants can, *potentially*, all be used up again by reaction with carbon compounds in the process of respiration by plants themselves. Of course when a plant dies it stops respiring, but what then happens to it? On the land surface, and in the upper waters of the ocean where there is oxygen available, dead animals and plants are used as food by other organisms which, when they respire, consume oxygen and release carbon dioxide. Bacteria are especially important organisms in the decay of dead matter and their respiration consumes a correspondingly large amount of oxygen.

Plants that are fully decomposed by bacteria and other organisms after they die therefore make no net contribution to the amount of oxygen in the atmosphere.

The only way oxygen could *accumulate* in the atmosphere would be if the dead remains of the phytoplankton were somehow locked away, where they couldn't react with oxygen. This is in fact what happened, and what still happens. Remains of phytoplankton, and the remains of the animals which eat them, sink to the bottom of the ocean. Large quantities of unrespired carbon sit on the bottom of the oceans, the residue of organisms to which we owe our oxygen-rich atmosphere.

Carbon derived from ancient photosynthesis is present in another kind of rock: coal. Coal is formed of the fossilised remains of undecomposed land plants. The carbon in these rocks was removed from the atmosphere by photosynthesis during late Palaeozoic times, and because the carbon was 'locked away' and not immediately released again by decomposing organisms, the oxygen produced by this photosynthesis became a net addition to that in the atmosphere.

Q   Approximately when did the Palaeozoic begin?

A   About 590×10^6 years ago (see Figure 5.2).

Q   Did the atmosphere contain oxygen before this time?

A   Yes. The first oxygen-producers were the stromatolites which appeared in the Precambrian. Red beds indicating an atmosphere containing oxygen also appeared in the Precambrian (Figure 5.2).

Photosynthesis by phytoplankton removes dissolved carbon dioxide from sea water, but a particular group of marine phytoplankton, the coccoliths, remove additional carbon dioxide from solution and form shells of calcium carbonate with it as well. The shells of these organisms deposited at the bottom of the ocean formed limestone. The chalk of the White Cliffs of Dover is an example of an especially pure limestone, laid down in the late Cretaceous, when it is estimated that the seas rose to cover 82% of the Earth's surface. The process of shell formation does not involve photosynthesis directly (although this is where the plankton acquire the energy for the process) and so it does not involve the liberation of oxygen. However, it does, indirectly, remove carbon dioxide from the atmosphere. Put simply, it works as follows. Carbon dioxide dissolves in water – thereby providing a sort of reservoir of 'dissolved' carbon that coccoliths (and some other marine organisms) can draw on to make calcium carbonate to build their shells. This would tend to drain the reservoir – removing carbon dioxide from solution and locking it up as solid calcium carbonate – were it not that the loss from sea water can be 'topped up' by further dissolution of carbon dioxide from the atmosphere. The overall effect is, then, to remove carbon dioxide from the air and to bury the carbon in skeletal remains at the bottom of the ocean.

Note that, unlike the formation of coal and the deposition of other phytoplankton remains, the formation of limestone rock did not produce an increase in atmospheric oxygen, but its contribution to the reduction of carbon dioxide concentration in the atmosphere through geological time was probably greater than that of photosynthesis.

Q   Approximately when did the Cretaceous begin?

A 145×10^6 years ago (Figure 5.2).

Q   Were any limestones laid down before this?

A   Yes. The earliest limestones date from the Precambrian (Figure 5.2).

Up to now we have only really considered one dimension of the geological record: time. When we look at the geographic distribution of sedimentary rocks around the globe, and at the fossils in them and the kinds of environments they lived in we encounter some very strange facts. The tallest mountain ranges of the world – the Himalayas, the Alps, the Andes, the Rockies – are largely composed of limestone laid down in the sea! But wait, there is worse! In Cumbria there are rocks from the southern hemisphere, in Antarctica we find deposits that were laid down in tropical environments and in tropical areas we find evidence of glaciation. The detailed explanations for such anomalous distributions are beyond the scope of this text, but many anomalies share the same explanation. This is at one and the same time devastatingly simple and simply devastating: **continental drift**.

2.4   Continental drift

The Earth's crust and upper mantle form a layer called the **lithosphere** (see Figure 5.1). The lithosphere is divided into segments or plates which, it is believed, move around the Earth with respect to one another. The age of the crust varies over the surface of the Earth. In continental areas the oldest rocks were formed about 3900 million years ago, whereas in oceanic areas the crust is nowhere older than 200 million years. Molten rock from the mantle (magma) surges to the surface in submarine volcanoes along mountain ridges running through all the oceans (Figure 5.3). Because new crust is formed along these ridges, they are known as constructive plate margins. Either side of these mountain ridges the sea floor travels sideways, moving away at rates of about a few centimetres (cm) a year in a process known as sea-floor spreading.

Continents are carried on their lithospheric plates away from constructive plate margins: for example, the continents of South America

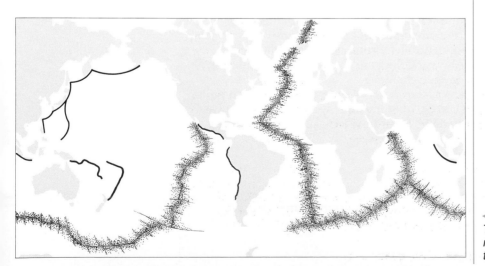

◀   Figure 5.3
The ocean floor showing ridges (shaded) and trenches (solid lines).

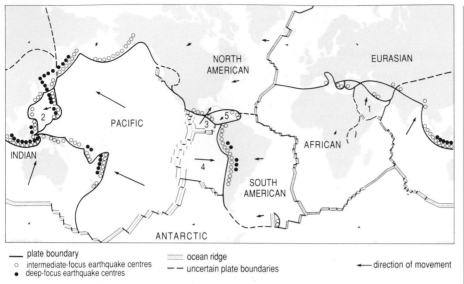

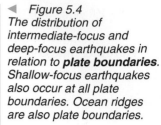

◄ *Figure 5.4*
The distribution of intermediate-focus and deep-focus earthquakes in relation to **plate boundaries**. *Shallow-focus earthquakes also occur at all plate boundaries. Ocean ridges are also plate boundaries.*

—— plate boundary   ⋯⋯ ocean ridge
○ intermediate-focus earthquake centres   — — uncertain plate boundaries   ◄——— direction of movement
● deep-focus earthquake centres

Minor plates: 1 Arabian; 2 Philippine; 3 Cocos; 4 Nazca; 5 Caribbean

and Africa are travelling away from each other, widening the South Atlantic Ocean at the moment.

As well as being older, the continental crust is five to six times thicker and is more buoyant than the oceanic crust. Where the thin, denser crust of the ocean floor reaches the continental margins, it may be forced down into the mantle along subduction zones which form oceanic trenches, such as the one which reaches a depth of 8 kilometres (km) below sea level along the Pacific coast of South America (see Figure 5.1).

The stresses in the mantle along subduction zones cause seismic activity. These zones are the locations of many large earthquakes (Figure 5.4) and nearby areas experience major volcanic activity. San Francisco, which was flattened by a major earthquake in 1906, damaged in 1984 and 1989, and is 'holding its breath' for the next big one, sits on the San Andreas fault where the American and Pacific plates slide past each other.

The Himalayas are geologically recent mountains and were thrown up by the impact of India with Eurasia when their respective plates collided. The theory of **plate tectonics** (the movement of plates) not only determines the geographical distribution of natural environmental hazards such as earthquakes, but also explains how the flora and fauna of certain regions of the world have attained their unique character.

Having looked at the oceans as the source of limestone rocks that build mountains, as a depository of carbon from the atmosphere, and indeed as the first environment of primeval life, let us now look at the ocean realm and the role it plays in the Earth of today.

Activity 1

Try to answer the following questions, which review some important points in Section 2. You will find answers and comments at the end of the chapter.

Q1   Name the three main types of rock, and how they are formed.

Q2   Which one of the following list is the unique most characteristic of living organisms?

A   Living organisms are made up of hydrogen, carbon and oxygen.

B   Living organisms obtain their energy from the reaction between oxygen and carbon compounds.

C   Living organisms are capable of movement.

D   Living organisms are capable of reproduction.

Questions 3–6   Substances essential to life include water, hydrogen, oxygen, carbon dioxide, carbon compounds, nitrogen.

Q3   Which of the substances in the list are consumed by respiration?

Q4   Which of the substances in the list are produced by respiration?

Q5   Which of the substances in the list are consumed by photosynthesis?

Q6   Which of the substances in the list are produced by photosynthesis?

Q7   In what important ways have living organisms changed their physical environment through time? What is the geological evidence for this?

Q8   Why do plants which die and are totally decomposed or burnt have no net effect on the amount of oxygen or carbon dioxide in the atmosphere?

Q9   In what *two* ways do phytoplankton which form shells of calcium carbonate remove carbon dioxide from solution in sea water?

2.5   Summary

At the time of the Earth's formation there were neither oceans nor atmosphere. The first atmosphere was formed from the outpouring of volcanoes and consisted mostly of carbon dioxide.

Geological events are measured on a time scale called the stratigraphic column. The long periods of time involved in Earth history, and large numbers in general, can be described using powers of ten (see Box 5.1). For example, one million is written 10^6; the Earth is about 4.6×10^9 years old.

During the Precambrian the atmosphere changed from a composition rich in carbon dioxide and devoid of oxygen, to a composition rich in oxygen and poorer in carbon dioxide. These changes are detectable in the chemical composition of Precambrian rocks, and coincide with the appearance of blue-green bacteria able to produce free oxygen by photosynthesis.

Photosynthesis occurs in all green plants and uses the energy of sunlight to turn carbon dioxide and water into carbon compounds and oxygen. Respiration is the chemical equivalent of combustion, and the reverse of photosynthesis. It occurs in all living things. If all the carbon compounds produced by plants were used up in respiration, either by the plants themselves, or by bacteria and animals feeding on them, then the oxygen generated by photosynthesis would all be converted to carbon dioxide. This has not happened because unrespired carbon compounds have become buried in ocean sediments where they are inaccessible.

Fossil fuels such as coal are another example of buried, unrespired carbon compounds. The formation of limestone from the shells of marine plants contributed to the removal of carbon dioxide from the atmosphere, but did not directly increase atmospheric oxygen.

The upper layer of the Earth, called the lithosphere, is divided into plates which move over its surface. This movement has caused continents to drift. The margins of plates are associated with earthquakes and volcanoes.

3   Atmosphere

3.1   Introduction

> For the first time in my life I saw the horizon as a curved line. It was
> accentuated by a thin seam of dark blue light – our atmosphere.
> Obviously this was not the ocean of air I had been told it was so many
> times in my life. I was terrified by its fragile appearance.

These words of Ulf Merbold, an astronaut, describing his first view of Earth from space (Kelly, 1988*), emphasise how vulnerable life on our planet is, protected and nourished by a thin film of gas. The atmosphere insulates the Earth from the daily extremes of temperature that would freeze the world by night and scorch it by day if, like the moon, the Earth had no air. As near as makes no difference, the moon and the Earth are the same distance from the sun, but the surface of the moon rises to 100 °C by its day and plummets to –150 °C by its night.

3.2   Atmospheric gases

By volume, the atmosphere comprises only two main gases. You should recall that these are nitrogen and oxygen (see Table 5.1). An inert gas called argon comprises most of the remaining 1%. The air contains only 0.03% carbon dioxide, and of course, water vapour in varying amounts, as well as some other gases.

Another gas which is in even lower concentration than carbon dioxide, but which is also very important to life, is **ozone**. Ozone high up in the atmosphere has beneficial effects, but these are nothing to do with what some people by the seaside call 'ozone', which isn't this gas at all, but the smell of rotting seaweed! Ozone is highly toxic if it comes into direct contact with life and is one of the pollutants in smog, which is produced near ground level by the action of sunlight on the gases from vehicle exhausts.

The biological importance of ozone higher up in the atmosphere is that it filters out some of the **ultra-violet (UV) light** from the sun, reducing the

*KELLY, K. W. (1988) *The Home Planet: images and reflections of earth from space explorers*, Macdonald, Queen Anne Press, London.

intensity which reaches the Earth's surface. Ultra-violet light, which is invisible to the human eye, causes sunburn and the tanning of skin exposed to sunlight. Skin pigmentation is a protection against UV which damages living tissue and, after severe exposure, can cause skin cancer.

Up to an altitude of about 80 km the atmospheric gases occur in roughly constant proportions, but water vapour and ozone are exceptions. In both cases, the dependence on altitude is related to the source of the gas. Water vapour comprises up to 4% of the atmosphere near the surface, but is almost absent above about 10–12 km. This reflects its source – largely evaporation from surface waters or evapotranspiration from plants (explained further in Section 5.2). From here, water vapour is transported upwards by atmospheric turbulence, which is most effective below about 10 km. This, combined with the fact that the maximum possible water vapour content of cold air is anyway very low, means there is little water vapour in the upper atmosphere.

By contrast, ozone reaches its maximum concentration at an altitude of 20–25 km. The concentration of ozone increases with altitude because it is formed from oxygen by chemical reactions high up that are powered by UV light. Sunlight is more intense here than at ground level because it has not been filtered by a large thickness of air, or by the **ozone layer** itself.

The temperature of the atmosphere changes with altitude in a rather complicated way. This is best shown by a graph on which temperature is plotted against height (Figure 5.5). If you are not familiar with the interpretation of graphs, you should now look at Box 5.2.

The explanation for the sinuous pattern of variation in temperature with altitude (Figure 5.5) is a complicated one, which we cannot enter into fully here. Essentially, there are two sources of heat which warm the atmosphere, one from below and one from above. At the surface of the Earth the ground receives light, and radiates this back into the atmosphere as heat. In the upper atmosphere some solar radiation is absorbed by gases before it reaches the ground, and this absorption of light warms the atmosphere too.

In the bottom-most layer of the atmosphere, called the **troposphere**, the warmth of the Earth is the predominant source of heat. Consequently air temperature drops with altitude as distance from the Earth's surface increases.

Q Using the graph in Figure 5.5, find the relationship between air temperature and altitude as you climb from sea-level to the top of a mountain.

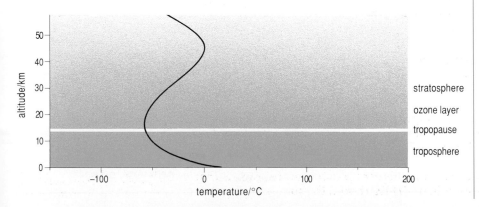

◄ Figure 5.5
The temperature profile of the atmosphere.

Box 5.2   How to read a simple graph

Graphs are designed to show how the quantity of one measurement changes with the quantity of another. In the case of Figure 5.5, the graph shows how temperature, measured in degrees centigrade (°C), changes with height (altitude) in the atmosphere, measured in kilometres (km). Quantities like 'temperature' and 'height', are described as the variables of the graph (because they vary!). In this example height is plotted up the side of the graph, on its vertical axis, and runs from zero (which represents sea-level) to over 50. Temperature is plotted along the bottom of the graph, on its horizontal axis, and runs from minus 100 °C to plus 200 °C. The ranges of numbers along each of the axes are called the scales, and each is expressed in the unit of measurement shown after the oblique stroke in the label on the axis: altitude/km, temperature/°C. The line running through the body of the graph is called the curve. In this particular case the curve *is* actually bent, but in other graphs where it might be a straight line it is still, by convention, called the 'curve'.

The graph allows you to take any value on the height scale and to say what the temperature at that height is. For example, if you want to know what the temperature is at 20 km altitude:

1   Draw a horizontal line on Figure 5.6 from the 20-km point on the vertical scale across the graph until it hits the curve.

2   Then, draw a vertical line from this point on the curve, downwards until it hits the horizontal axis.

3   Read off the temperature on the horizontal scale, at the point where your vertical line hits the horizontal axis.

Q   Do this now, and write down the answer you get.

A   You should have got the answer −55 °C.

If you want to know the height (or heights) at which the temperature of the atmosphere measures −20 °C, you can do this in the same way, but start from −20 °C on the horizontal axis and read off the answer on the scale of the vertical axis.

Q   Do this now, and write down the answer(s) you get.

A   You should have obtained three answers to this question, because a vertical line from −20 °C hits the curve at three places, corresponding to three different heights. The three answers are 4 km, 35 km and 55 km.

Graphs can be used in the way you have just practised to give a value for the temperature at a particular height (or vice versa), but they are most useful for portraying the general pattern of a relationship between two variables. So, in the case of Figure 5.5, what is most interesting is not so much that the temperature is about −20 °C at a height of 4 km, but that as altitude rises, the temperature first decreases, then abruptly begins to increase again, and that the curve switches direction again higher up.

Now, return to the main text and we will see the significance of this pattern.

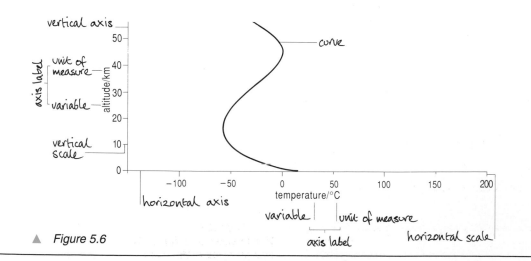

▲   Figure 5.6

A   Everest, the world's tallest mountain, is 8.85 km tall, which is well within the region where temperature falls with altitude, as shown in the graph in Figure 5.5.

In general, air temperature falls by an average of about 6.5 °C with every 1000 m you ascend. Note however, that there are certain weather conditions which can greatly alter this, and even cause temperature inversions near the ground, so that a layer of warm air sits on top of a layer of cold air.

At an altitude of about 10 km the warming of the atmosphere by the absorption of solar radiation begins to become significant. At this point, called the tropopause, the relationship between temperature and altitude changes direction and air temperature begins to *rise* with altitude. The absorption of UV by ozone and the warming of the air which results is chiefly responsible for the increase in temperature with altitude which occurs above the tropopause, in the region called the **stratosphere** (see Figure 5.5).

The altitude of the tropopause varies somewhat depending upon latitude and weather conditions. It acts as a kind of lid on the troposphere, limiting the mixing between the atmospheric layers above and below it.

The thermal insulating properties of the Earth's atmosphere are largely due to water vapour and carbon dioxide contained in the troposphere. Water vapour and carbon dioxide are more transparent to radiation from the sun than to the heat which the Earth radiates when its surface is warmed (Figure 5.7). Carbon dioxide and water vapour are called **greenhouse gases** by analogy with a glasshouse which traps the heat from solar radiation. Three other important greenhouse gases are ozone, methane (which is produced by bacteria living in oxygen-less conditions, e.g. the stomachs of cows), and nitrous oxide which is produced by soil bacteria. Synthetic compounds called chlorofluorocarbons (CFCs) which are manufactured for use in refrigerators and other cooling systems such as air conditioners also behave like greenhouse gases. They are also used as

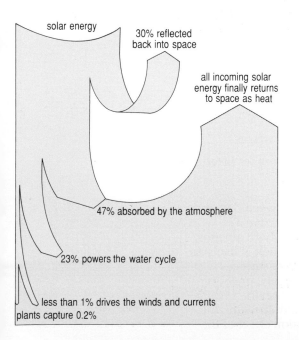

solar energy

30% reflected back into space

all incoming solar energy finally returns to space as heat

47% absorbed by the atmosphere

23% powers the water cycle

less than 1% drives the winds and currents

plants capture 0.2%

◄   *Figure 5.7*
What happens to the sun's energy in the atmosphere.

solvents, especially in the electronics industry, and as propellants in some aerosol cans. Some CFCs have a greenhouse effect 10^4 times as great as an equivalent quantity of carbon dioxide.

Methane is present in only very small amounts in the atmosphere, though this quantity is also increasing. CFCs are also present in trace amounts, but they have already reached concentrations that are high enough to damage the protective properties of the atmosphere. At the moment, their most damaging role is to speed up chemical reactions in the stratosphere which destroy ozone, and there is now good evidence that they are responsible for a massive depletion of ozone in the atmosphere over the Antarctic during the southern hemisphere spring, creating a seasonal ozone hole. Changes in atmospheric composition due to human activities are examined in greater detail later in the series.

Q   What biological effects are to be feared from damage to the ozone layer?

A   More UV radiation would reach the surface of the Earth, which could lead to an increase in skin cancers. Severe UV exposure also causes cataracts, and may damage the immune system.

Ozone is produced from oxygen by chemical reactions in the stratosphere powered by UV light. Therefore, although oxygen is a direct product of biological activity, ozone is an indirect one. It is created, performs its protective function and is destroyed in an atmospheric layer with no direct contact with living organisms. By contrast, carbon dioxide is directly and intimately involved with every living organism on Earth.

3.3   *Ice, clouds and temperature*

The overall **temperature of the Earth** is the result of a balance between the rate at which energy comes in from the sun and the rate at which it is radiated out again. If the two rates fail to match, the Earth will cool down or heat up until a balance is restored.

The proportion of radiation that a surface reflects is called its albedo. A surface with a high albedo, such as a mirror, appears bright to the observer looking at it, and conversely a surface with a low albedo such as a school blackboard appears dark. The albedo of clouds and other surfaces is shown in Table 5.2.

Q   How does the albedo of fresh snow compare with other surfaces?

A   Fresh snow has a higher albedo than water or any land surfaces listed in Table 5.2.

If you have ever experienced the extremes of temperature between day and night in a ski resort in winter, you know for yourself how effective a reflector of sunlight snow is. Because snow and ice have such high albedos, an increase in temperature that is sufficient to cause them to melt is likely to result in a further increase in temperature as the less reflective surfaces beneath them are exposed to sunlight and warm up. Conversely, once an extensive ice cap begins to form it will cause a further reduction in temperature because up to 90% of the incident radiation from the sun will be reflected from the snow-covered surface. The amplification effect of the

Table 5.2   The average albedo of Earth surfaces

Surface	Albedo/%
clouds	
cumulonimbus	90
stratocumulus	60
cirrus	40–50
fresh snow	80–90
water	6–10
sand	30–35
forest	7–18
grass and cereal crops	18–25

Source: Barry, R. G. & Chorley, R. J. (1987) *Atmosphere, Weather and Climate*, 5th edn, Methuen, London, Table 1.3, p. 19.

polar ice caps is an example of a process called positive feedback and can work in either direction, accelerating warming or cooling, depending upon initial temperature.

Q   Apart from fresh snow, which surfaces shown in Table 5.2 have high albedos?

A   Clouds, particularly the very thick, tall kind called cumulonimbus. These have an albedo which is as reflective as snow.

Clouds are also involved in a feedback process which affects the energy balance of the Earth, but this works in a rather different way, tending to cancel out rather than amplify changes. Averaged over the Earth's surface, clouds reflect back into space 21% of the total solar radiation reaching the planet, before it even gets to the ground.

Q   How would you expect a decrease in cloud cover over the oceans to affect the energy balance of the Earth?

A   Clouds have an albedo that is up to nine times greater than a water surface, so that a small decrease in cloud cover could lead to a large increase in the amount of solar radiation absorbed by the ocean.

Clearly, because the ocean and most land surfaces have a lower albedo than clouds, the amount of cloud over the Earth is likely to affect its temperature. What controls the amount of cloud over the ocean? This is actually quite a complicated question, but an important factor is the moisture content of the air. When the sky is clear of clouds, the surface of the ocean absorbs solar radiation which increases the temperature of the water. This, of course, causes an increase in the rate of evaporation of water vapour into the atmosphere. As the warmed air rises it expands and cools, and water vapour condenses on tiny windborne particles in the air, forming clouds. These clouds then reflect solar radiation so that less reaches the Earth's surface; the surface becomes cooler and evaporation decreases. Because evaporation increases when cloud cover is low, and decreases when it is high, this process tends to stabilise temperatures, like a thermostat. Mechanisms which cancel out changes in this way are said to work by **negative feedback**.

3.4   *Atmospheric circulation and climatic zones*

So far, the section has looked at the vertical structure of the atmosphere and its influence on solar radiation and temperature. To relate this to climatic zones, it is necessary to look at the global pattern of horizontal movement of the atmosphere and its consequences for climate. In detail, this is extremely complex, so in this short presentation only the basic outlines can be given. Four sets of factors influence the outcome: the relationship of the Earth's orbit to the sun; the distribution of the continents and oceans; the effects of gases and clouds in absorbing or reflecting incoming radiation and the albedo of the surfaces. Two of these have already been covered so this section will focus on the effects of the Earth's orbit and the effects of land and water.

It is obvious from our everyday experience that the power of the sun is greatest, other things being equal, when it is highest in the sky. In everyday life we experience this as the difference between midday and evening and between the sunshine of a clear December day and that of June. The seasonal variation between summer and winter depends on the changing tilt of the Earth towards the sun. Although the Earth revolves around the sun, the axis of the Earth's own rotation is not perpendicular to the plane of revolution, being inclined to it at an angle of $23\frac{1}{2}$ degrees. The result of this, as shown in Figure 5.8, is that in June the northern hemisphere is tilted towards the sun, which seems higher in the sky and gives longer days, whereas January sees short days and a low sun (this effect is reversed in the southern hemisphere). During the course of the year, the sun appears overhead at the equator on March 21, 'moves' north to $23\frac{1}{2}°$ N on June 21, 'returns' to the equator by September 23, and 'moves' to $23\frac{1}{2}°$ S by December 22. The result is that while the sun is generally high in the sky near the equator and low in the sky near the poles, there are considerable seasonal differences, which are particularly extreme in the latitudes near the poles.

In spite of the apparent north–south movement of the sun, its effectiveness at high angles is much greater than its effectiveness at low angles so that because of the curvature of the Earth the majority of solar energy is received near the equator. The equatorial areas would be much

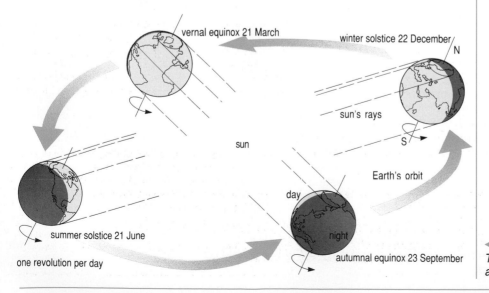

◄ *Figure 5.8*
The Earth's orbit
and the seasons.

hotter and the poles much colder if the atmosphere and oceans did not carry heat towards the poles. The basic mechanism is the formation of a gigantic convection cell where air at the equator is heated and hence rises, pulling in air along the surface from both north and south (Figure 5.9). This air, being warm, absorbs moisture from both land and sea. As this moist air rises, it cools and produces heavy rain, making equatorial areas very wet. The condensation process gives out more heat, strengthening the tendency to rise. As this rising air current nears the tropopause, its vertical movement is slowed and it moves horizontally both north and south. The convection cell is completed by the air sinking in subtropical latitudes before being pulled back towards the equator again. Because it is dry and sinking this air does not form clouds and is responsible for the clear skies of the hot desert belts.

A complication of this convection cell, where surface air flows toward the equator and high altitude air flows away from it, is that the rotation of the Earth deflects the apparent movement of the air to make surface winds follow a diagonal pattern, as shown in Figure 5.10.

Outside the tropical zone air circulation is more variable and more complex, influenced both by seasonal fluctuations in solar radiation and by the different properties of land masses and oceans. The crucial difference is that the continents heat up much more than the seas in summer but cool much more in winter. The summer heating of the continents causes the air to rise, drawing in surface winds to create smaller convection systems, once again with the rotation of the Earth producing a deflection so that winds blow diagonally around air masses. In winter, cooling of the continents tends to reverse this convection pattern, with air sinking over the continents and rising over the oceans, but the whole system is less

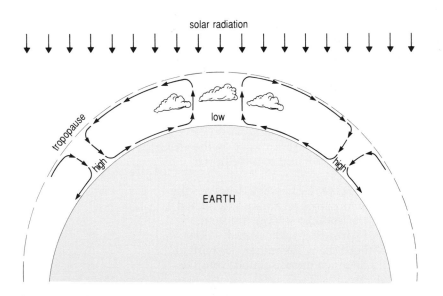

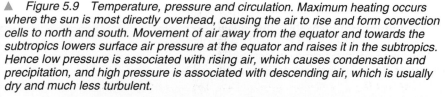

▲ *Figure 5.9   Temperature, pressure and circulation. Maximum heating occurs where the sun is most directly overhead, causing the air to rise and form convection cells to north and south. Movement of air away from the equator and towards the subtropics lowers surface air pressure at the equator and raises it in the subtropics. Hence low pressure is associated with rising air, which causes condensation and precipitation, and high pressure is associated with descending air, which is usually dry and much less turbulent.*

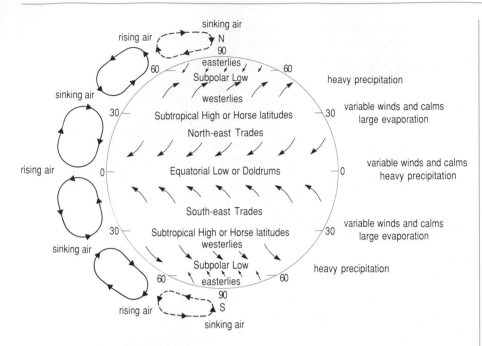

▲   *Figure 5.10   Atmospheric circulation.*

energetic. Because the southern hemisphere has a lower proportion of land
mass, these continental effects are less developed. The subtropical high
pressure belt is more continuous and the mid-latitude belt of westerly
winds is more constant and more energetic, as shown by the nickname 'the
roaring forties'.

     In higher latitudes, including that of the UK, the variability of weather
is a result of the opposition between westerly winds and colder air masses
pushing down from the poles. Because the westerlies are warmer and
wetter than the polar air, they tend to rise above it, cooling and condensing
to produce cloud, rain and sometimes snow. As in the tropics, condensation
releases heat and so encourages further rising. Because of the importance of
the westerlies at these latitudes, oceanic influences (relatively constant
temperatures and higher humidity) are brought into coastal areas on the
west side of continents, while continental influences are stronger on the east
side. Hence, east coasts of continents tend to be drier and to have hotter
summers and colder winters than the west sides.

     A much less obvious mechanism which helps to carry heat towards the
poles occurs where the upper air westerlies form into jet streams, which at
their fastest exceed 400 km per hour (k.p.h.). These jet streams fluctuate
north and south and are a potent means of spreading warmer and wetter
air into polar regions at high altitude. Without them, the rather stagnant air
masses of the Arctic and Antarctic would be much more isolated from the
more active circulation systems than they are.

     To sum up, the broad pattern of circulation is composed of stable
systems in tropical and polar regions, both typified by easterly flows at the
surface, and an unstable zone in the middle latitudes, showing variations
between seasons, between land and ocean and between tropical and polar
air. This circulation creates the patterns of climatic zones which will be
discussed in Chapter 6. It is also one of the chief influences on the
circulation of the oceans.

Activity 2

Try to answer the following questions, which review some important points in Section 3.

Q1 In which layer of the atmosphere is: (a) most water vapour to be found? (b) the ozone layer?

Q2 Why does temperature fall with altitude in the troposphere, and rise with altitude in the stratosphere?

Q3 Name three naturally occurring greenhouse gases. What properties do these gases have which earn them this name?

Q4 From what other gas is ozone in the ozone layer formed, and how does its presence there protect living things on the Earth's surface?

Q5 What is meant by the albedo of a surface? Why is this characteristic important to climate?

Q6 How can the ice cap amplify changes in the Earth's temperature?

Q7 In what way can clouds operate like a thermostatic control on the temperature of the Earth?

Q8 In which two of the four main climatic belts (tropical, low, subtropical high, mid-latitude low, polar high) are clouds more common, and what effect does this have on receipt of solar radiation?

3.5   Summary

The atmosphere is a thin film of gas enveloping the Earth. It acts as a blanket, raising and stabilising the surface temperature. The bottom-most layer of the atmosphere is called the troposphere and is warmed by heat from the Earth's surface. Within this layer, air temperature falls with altitude as the source of heat recedes. The layer above the troposphere is called the stratosphere, and here air temperature rises with altitude as progressively more of the sun's energy is absorbed by the atmosphere. Ozone, which occurs in the stratosphere in a layer at 20–25 km altitude, is particularly important because it absorbs ultra-violet light, reducing the intensity of this radiation at the Earth's surface where it is harmful to life.

Carbon dioxide, water vapour, methane, ozone and nitrous oxide are the gases largely responsible for the atmosphere's insulating properties and are called greenhouse gases. CFCs are synthetic gases, some of which have strong greenhouse properties. They are also important because they destroy ozone in the stratosphere.

The temperature of the Earth is determined by a balance between the rate at which energy comes in from the sun and is radiated out again. The albedo of various types of surface have a strong influence upon this. Ice has a high albedo and can have a destabilising effect through positive feedback between temperature and the area of the ice cap. Clouds also have a high albedo, but have a stabilising effect upon temperature through negative feedback between cloud formation and temperature.

The greater part of solar energy is received in equatorial and tropical latitudes. The apparent movement of the sun north and south of the equator produces the seasons. Solar heating produces a large-scale convection system with low pressure near the equator and high pressure at about 30° N and 30° S. During the summer these high pressure belts are interrupted by low pressure areas caused by continental heating, especially in the larger northern continents. The mid-latitudes are on balance areas of low pressure, but are areas of variable climate as a result of seasonal fluctuations, differences between continents and oceans and alternation of tropical and polar air masses. The polar areas are characterised by low temperatures and high pressure. The net effect of the circulation of the atmosphere is to transfer heat from the tropics to the mid-latitudes, making surface climates more equal than they would be if the atmosphere were static. This transfer is further assisted by the oceans.

4   The oceans

4.1   Introduction

In effect, the oceans are a huge reservoir formed by the interconnected basins of the Atlantic, Pacific, Antarctic and Indian Oceans. This reservoir holds not only water, but also heat and dissolved atmospheric gases (mainly nitrogen, oxygen and carbon dioxide), and of course salt. The movement of water, heat and gases between the ocean and atmosphere balances the global climate, though how this happens is still not well understood.

Biologically, the oceans make a major contribution to the fixation of carbon dioxide by the photosynthesis of phytoplankton. These tiny plants are estimated to fix 16×10^9 tonnes of carbon each year. This is the basis for virtually all marine life, including the annual harvest of fish which we take from the sea for food and fertiliser.

The level of the oceans with respect to the land has varied greatly through geological time.

Q   During which geological period mentioned in Section 2 was there a large rise in sea-level which influenced the geology of Europe today?

A   The Late Cretaceous, when the chalk was laid down.

Such large changes in sea-level are thought to have been associated with continental drift. Smaller, but important, changes in sea-level occur with changes in global temperature.

Q   The density of sea water decreases when temperature increases. As its density decreases water expands, so how would an increase in mean global temperature affect average sea-level?

A   It would raise sea-level.

An expansion in the volume occupied by water in the oceans as temperature rises is called thermal expansion.

A change in global temperature also has another effect on sea-level because, among other things, it affects the relative amount of water that is stored in the polar ice caps.

Q Ice is actually less dense than water, which is why it floats. What effect would the melting of icebergs have on sea-level?

A None at all, because floating icebergs displace a volume of water of weight equal to their own weight. When the ice melts, its density increases but its weight remains the same, so the volume of water it produces is equal to that it previously displaced.

Activity 3

You can verify that melting icebergs don't raise sea-level for yourself by performing the following experiment.
   Fill a glass of water containing an ice cube to the brim, and place it on a plate or saucer. Does the glass overflow when the ice melts?

The ice which covers the Arctic Ocean is floating, but ice deposited on land, such as in Antarctica, would cause a rise in sea-level when it melted and entered the ocean. During glacial periods in the geological epoch known as the Pleistocene (between 10 000 and 2 million years ago), there were falls in sea-level, on occasions by more than 100 m, caused by accumulation of ice on land. In the interglacials sea-levels rose. Such a rise in sea-level cut Britain off from the European continent about 8000 years ago, and in the process prevented the recolonisation of the British Isles by many plants and animals that are now found just the other side of the English Channel.
   A relatively small rise in sea-level today would endanger most of the world's major cities. Hence there is cause for concern that global mean temperature appears to have risen by about 0.5 °C in the last 100 years and mean sea-level has increased by 100–150 mm over the same period. Most of this rise is thought to be due to the thermal expansion of sea water. Some scientists believe that the rise in global mean temperature which has caused the rise in sea-level is related to changes in the composition of the atmosphere, and the greenhouse effect which we discussed in Section 3.

4.2   Ocean currents and marine life

Ocean currents have a significant effect on the geographical distribution of the ocean's biological resources and upon the ocean's climatic effects (Figure 5.11). The major surface currents are driven by winds, but several other forces are also involved, in particular the high pressure areas in northern and southern ocean basins and the associated wind systems. As a consequence, the major ocean currents travel clockwise in the ocean basins of the northern hemisphere and anticlockwise in the southern oceans. The Gulf Stream is a part of the circulatory current in the North Atlantic from which the North Atlantic Drift is an extension that warms the shores of northern Europe.
   A general physical property of gases and liquids is that they rise when warmed and sink when cooled. In the deep waters of the open ocean the water tends to stratify into a warm surface layer and a cold bottom layer, separated by a temperature gradient called a thermocline. In the temperate

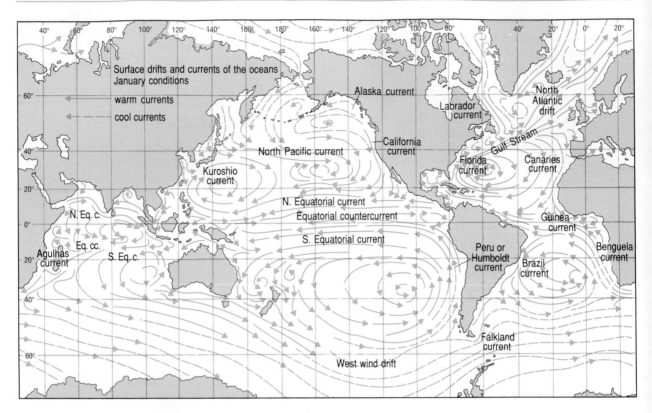

zone the thermocline breaks down in autumn as the surface waters cool and mix, but in the tropics where seasonal changes in temperature are less extreme, the temperature difference between the two layers is a stable one and they do not mix. As a consequence, tropical waters are in general impoverished because particles carrying nutrients essential to the growth of animals and plants sink to the bottom and remain there, out of reach of phytoplankton which can only grow near the surface where there is light.

Where prevailing winds blow from land to sea in the tropics, such as off the coast of Peru, these drag up deep, cold water laden with nutrients that create a bonanza for marine life. The nutrients in the upwelling water support phytoplankton which are fed upon by microscopic marine animals called zooplankton. These in their turn feed fish. Sea birds and humans depend upon the bounty of fish from this upwelling which is especially important because the neighbouring coast is an arid desert supporting little life.

The upwelling off the coast of Peru is prone to disruption by a warm current called *El Niño* that cuts off the supply of nutrients about one year in seven, with widespread catastrophic results. In fact it is now believed that the *El Niño* current is linked to atmospheric conditions that have widespread climatic repercussions in both the northern and southern hemispheres.

Another biologically very important region of high productivity occurs in polar waters around the Antarctic continent where the wind patterns cause surface waters to move in opposing directions. In this region, known as the Antarctic Divergence, upwelling cold water brings nutrients from depth which feed phytoplankton, which in turn are the food for small crustaceans (the group to which prawns belong) known as krill. These teeming zooplankton are the staple food of a large part of the marine life of

Antarctica, including penguins and whales. Krill are also fished in increasing quantity by trawlers, with unknown effects upon populations of marine animals which depend upon krill.

Activity 4

Try to answer the following questions, which review some important points in Section 4.

Q1   Name *two* ways in which an increase in global temperature can cause a rise in sea-level.

Q2   Describe the link between solar radiation, winds, and the major ocean currents.

Q3   Why is there rich marine life in ocean areas where there are upwellings of cold water?

4.3   Summary

The oceans are a reservoir of heat, atmospheric gases and salt as well as water. Sea-level has varied greatly over geological time, and is influenced by global temperature. A rise in temperature causes a rise in sea-level due to the thermal expansion of sea water and, if the temperature rise is great enough, by melting ice in the polar ice cap. When floating ice melts it does not affect sea-level.

Biologically the oceans make an important contribution to the fixation of carbon dioxide and they teem with life, particularly in regions enriched with nutrients by upwelling water, or the movement of ocean currents. The major ocean currents themselves are driven by the winds.

---

5   The water cycle

5.1   Introduction

The energy from the sun drives the winds, powers the ocean currents and supports all plant and animal life on the planet. Nearly all of this energy is eventually radiated back into space as heat, so it is convenient to think of energy as if it *flows* through the Earth system. However, all the materials involved in the machinery of living organisms, the oceans and the atmosphere neither enter nor leave the Earth, but are *cycled*. This is easiest to see for water. The **water cycle** is powered by the sun. In fact, over half of the sun's energy that reaches the Earth's surface is used up in evaporation. We can recover some of this energy as hydro-electric power from rivers as they make their way to the sea.

Each depository of water, such as the polar ice caps, lake water, the oceans and so on can be regarded as a compartment in the water cycle. The volume of water in each compartment and the rate of flow between them as far as we understand it at present are shown in Figure 5.12.

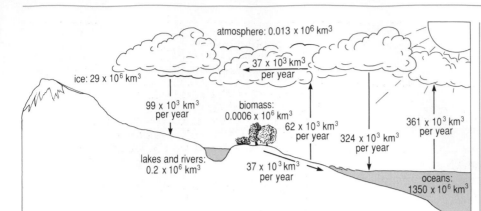

◀ *Figure 5.12*
The water cycle.

5.2   *Volumes of water and rates of flow*

The usual unit for measuring volumes is the litre, but when it comes to measuring the volume of water in the oceans, a bigger yardstick is needed. The one we use for this purpose is the cubic kilometre. A cubic kilometre is 1 km wide×1 km long×1 km high, so we use a superscript 3 to indicate that we are measuring in kilometres cubed: km^3. All the numbers below that refer to Figure 5.12 are measured in this unit. You already know how to convert large numbers into powers of ten, and you will be using this below. One further useful trick to know about using such numbers is how to add, subtract, multiply and divide them. Now study Box 5.3, which tells you how to do this.

When you have studied Box 5.3, you are ready to deal with the large quantities of water involved in the global cycle.

Q   Using the information in Figure 5.12, calculate the total volume of water in the five compartments of the water cycle.

A   Listing the quantities from the figure, you should obtain the answer shown in Table 5.3.

All the values shown in the figure and the table are *estimates*, because it is impossible to make actual measurements of global quantities like these. When you see a table of numbers like this you should always ask yourself how accurate they are likely to be, or in other words, 'How near the real

Table 5.3   Estimates of the quantities of water in various compartments of the water cycle, taken from Figure 5.12

Compartment	Volume of water/(10^6 km³)
oceans	1350
lakes and rivers	0.2
ice	29
organic matter	0.0006
atmosphere	0.013
Total	1379.2136

Box 5.3   Calculations with powers of ten

To add powers of ten

First convert the numbers to a form in which both have the same power.

For example, to add (1×10^2) to (3×10^5), you would rewrite (3×10^5) as (3000×10^2), ie. $(3 \times 10 \times 10 \times 10 \times 10^2)$.

Then the two numbers can simply be added by summing their left-hand sides:

1+3000 = 3001

and then putting back the power to give you 3001×10^2.

If you wish, you can then rewrite 3001×10^2 as 3.001×10^5.

To subtract powers of ten

To subtract two numbers, convert them to a form in which both have the same power (just as before), and then subtract one left-hand side from the other. For example,

$(5.1 \times 10^{10}) - (1.3 \times 10^9)$

is the same as

$(51 \times 10^9) - (1.3 \times 10^9)$

Subtract the two left-hand sides

51 − 1.3 = 49.7

and then put back the power to give you

49.7×10^9

If you wish to express your answer in the original form of the larger number, convert it:

$$49.7 \times 10^9 = \left(\frac{49.7}{10} \right) \times 10^{10} = 4.97 \times 10^{10}$$

To multiply powers of ten

When two numbers are mutiplied together, you multiply the two left-hand sides by one another and add the powers to each other, thus:

$(2 \times 10^6) \times (2 \times 10^4) = (2 \times 2) \times (10^{6+4}) = 4 \times 10^{10}$

To divide powers of ten

Conversely, when you divide one number by another, you divide the left-hand sides of one by the other, and subtract the powers from each other, thus:

$$\frac{2 \times 10^6}{2 \times 10^4} = \frac{2}{2} \times (10^{6-4}) = 1 \times 10^2$$

Now try these questions:

Q   A kilometre (written km) is 1×10^3 metres. How many cubic metres (written m³) are there in 1 km³?

A   1 km³ = 1 km × 1 km × 1 km

Remember that 1 km = 1×10^3 m, so

1 km³ = $(1 \times 10^3$ m$) \times (1 \times 10^3$ m$) \times (1 \times 10^3$ m$)$

The rule for multiplying powers of ten is to multiply the left-hand sides, and to *add* the powers, so:

1 km³ = $(1 \times 1 \times 1) \times (10^{3+3+3})$ m³

In 1 km³ there are 1×10^9 m³.

Notice that the answer is expressed in units of metres *cubed* (m³) because we have multiplied together three quantities, each expressed in metre units:

m × m × m = m³

---

values are these estimates?' This is not an easy question to answer, but there is a general point to be made here which applies to all estimates that compare or add up widely different quantities.

Notice the huge difference between the smallest and the largest values in Table 5.3. It is estimated that there are about two and a half million times as much water in the oceans as in organic matter. When all the values are expressed in the same units of measurement – units of millions of cubic kilometres (10^6 km³) – you can see that the estimate for the water in the oceans looks as though it has been made to the nearest ten, whereas the estimate for organic matter looks as though it has been made to the nearest ten-thousandth (0.0001). When we add up estimates that differ this much, and write down a total like 1379.2136, we give the erroneous impression that we know the answer to the nearest ten-thousandth of a unit. Of course this cannot be true, because by far the biggest part of the sum is the estimate for the amount of water in the ocean, and we have only estimated this to the nearest ten.

Bearing this point in mind, you can still compare the quantities shown in Figure 5.12, but without taking the answers as literal truth.

Q What percentage of the total water is to be found in the oceans?

A $\dfrac{(1350\times10^6)\ \text{km}^3}{(1379.2\times10^6)\ \text{km}^3}\times100 = 97.9\%$

Only 2% of all water is fresh, and 75% of this is locked up in glaciers and ice sheets. Most of the remainder is in groundwater (water in the subsoil), with only 0.33% of all freshwater held in lakes and rivers at any one time. Because there is so much more water in the oceans than anywhere else, these percentages wouldn't change very much, even if some of the other estimates in Figure 5.12 were out by as much as 100%.

Q What are the total amounts of water entering the atmosphere and the total amounts leaving it each year?

A The amounts, all in units of 10^3 km^3, are:

Entering from		Leaving to	
oceans	361	ocean	324
land	62	land	99
Totals	423		423

In other words, the *average* inputs and outputs of water to the atmosphere balance. If they did not, we could expect water to build up and up in the atmosphere, or the atmosphere to dry out completely. Because we know that this doesn't happen, the estimates have been *made* to balance by adjusting them.

Q Considering that 423×10^3 km^3 water passes through the atmospheric compartment each year and the quantity of water it holds at any one time (see Figure 5.12), how long on average does water spend in the atmosphere?

A If 423×10^3 km^3 passes through in a year, 1 km^3 passes through the atmosphere in

$$\frac{1}{423\times10^3}\ \text{years}$$

At any one time the atmosphere contains 0.013×10^6 km^3, which is the same as 13×10^3 km^3.

This passes through in

$$\frac{1}{423\times10^3}\times13\times10^3\ \text{years}$$

which is

$$\frac{13\times10^3}{423\times10^3}\times365\ \text{days} = 11.2\ \text{days}$$

Remember that the figure of 11.2 days has been calculated from a set of estimates made to different levels of accuracy. For this reason, more important than the figure of 11.2 is the unit, which is *days*. Whether the true figure is half 11.2, or half as much again, it is still a short time. The time you have calculated is known as the **residence time**. You can calculate the residence time of any substance that moves between compartments with the general formula:

$$\text{residence time} = \frac{\text{size of compartment}}{\text{rate of input (or output)}}$$

As you can see, the atmosphere holds very little water, and such water as it does hold has a very brief residence. This makes atmospheric moisture content very sensitive to variations in the size of inputs and outputs to the compartment, and is one of the reasons why rainfall can be so unpredictable.

Q What is the residence time of water in the ocean?

A Divide the water in the ocean by the annual input (or output):

$$\frac{(1350 \times 10^6)\ \text{km}^3}{(361 \times 10^3)\ \text{km}^3\ \text{per year}} = 3.74 \times 10^3\ \text{years}$$

With a residence time of nearly four thousand years, it is obvious that the amount of water in the ocean is likely to be little affected by small variations in the amount of evaporation from it.

Q Why does it follow from the long residence time of water in the ocean that a plan that was once put forward to prevent a rapid sea-level rise by trapping water on land behind dams wouldn't work?

A Because of the long residence time of water in the ocean, rainfall represents literally 'a drop in the ocean': very little of this could be captured for storage, and it would take a very long time to accumulate a significant amount of the ocean's water on land from that transferred there in rainfall.

Under 30% of the Earth's surface is land, and much of this is mountainous or arid. Even a dam (or dams) creating a lake that covered 10% of the Earth's land surface would only trap, on average, 3% of the rainfall each year. Even if this much could be caught, a lot would leak away and evaporate back into the water cycle.

The quantities given in Figure 5.12 are averaged over time and over the entire globe, which hides important climatic differences between one area and another. Some of these will be considered in a later chapter. One factor which affects the rate at which moisture enters the air from land and which is of global importance is the presence of plants, particularly trees. Plants lose water from their leaves by a process known as **transpiration**. Although only a small fraction of a leaf surface bears the pores through which moisture evaporates, the potential rate of water loss from a many-leaved canopy of leaves can approach that of loss from the same area of an open water surface. The total loss of water vapour from plants and soil is called **evapotranspiration**. The contribution trees make to this by transpiration can totally alter the rate of transfer of water between land and atmosphere. There are many reasons for being concerned about the environmental consequences of deforestation, but one which is particularly significant in the tropics is that it may change rainfall patterns on a regional scale. Rainfall, which would normally be trapped by tree roots and returned to the atmosphere through evapotranspiration, may simply run off into the sea if trees are removed. This will often cause soil erosion and loss of soil fertility too.

Activity 5

Try to answer the following questions, which review some important points in Section 5.

Questions 1–4 involve calculations with the following numbers expressed as powers of ten: (a) 1.5×10^3; (b) 1×10^6; (c) 4.8×10^9; (d) 3×10^2.

Q1 What is a+b, expressed in millions?

Q2 What is a–d, expressed in hundreds?

Q3 What is b×c?

Q4 What is a/d?

Q5 Which of the following numbers is nearest the residence time of water in the atmosphere which we calculated earlier?

  A 10^6 years   **D** 3.65×10^2 days

  B 10^3 years   **E** 10 days

  C 10 years

Q6 Name two different ways in which trees alter the flow of water in the water cycle.

5.3   Summary

The water cycle traces the movement of water between the oceans, the atmosphere, and the land surface where it occurs in ice, lakes and rivers and in living and dead matter. Over half of the solar energy reaching the Earth's surface is used up in the evaporation of water, which powers the water cycle. The oceans are by far the largest compartment of the water cycle. Calculations of the huge quantities involved in the water cycle illustrate how such large numbers can be dealt with, and why we must be cautious about the accuracy of estimates.

The residence time of water in the atmosphere is very brief. Evapotranspiration of water from plants, particularly trees, can strongly influence the rate at which water is transferred from land to atmosphere. Plants also influence the water cycle by trapping water among their roots, preventing run-off which may carry away soil as well as water.

6   The carbon cycle

6.1   Introduction

As we have seen, the evolution of life and photosynthesis by plants have changed the composition of the atmosphere during the Earth's history.

Q What three gases have changed in concentration as a result of biological processes? How have their concentrations changed?

A   Carbon dioxide, oxygen and ozone. The atmospheric concentration of carbon dioxide has fallen, whereas that of oxygen and ozone has increased.

Carbon, like water, passes through different compartments and this process is known as the **carbon cycle**.

Q   Which are the major compartments that you have already read about through which carbon passes, or in which it is stored?

A   Rocks (coal, limestone), organic matter (living and dead), the ocean (in solution), and the atmosphere.

Activity 6

Study Figure 5.13 and match each of the following processes with labels a to f in the figure that indicate *flows* of carbon between reservoirs:

- combustion of fossil fuels;
- deposition of calcium carbonate;
- photosynthesis by plants;
- respiration by organisms in the ocean;
- respiration by organisms on land;
- solution of carbon dioxide in sea water.

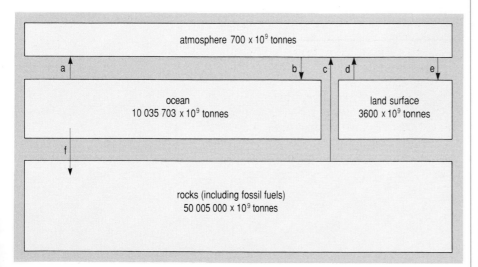

◄ *Figure 5.13*
Four major reservoirs of the carbon cycle. Arrows indicate fluxes of carbon between reservoirs (see text).

6.2 Carbon flows

Human activities affect several of the flows in the carbon cycle directly, and others indirectly. One of the direct effects is through deforestation. When plant growth, which consumes the carbon dioxide taken from the atmosphere by photosynthesis, is in balance with respiration and the decay of dead plants which releases it again, vegetation has no overall effect upon atmospheric carbon dioxide concentration. Burning down trees and not allowing them to regrow disturbs this balance, and results in the transfer of

carbon held in wood and other plant tissues to the atmosphere. The
burning of fossil fuel also increases the carbon flow to the atmosphere.

Q   What effect might you expect burning fossil fuel and extensive
    deforestation to have upon the concentration of atmospheric carbon
    dioxide?

A   Burning and extensive deforestation can both increase atmospheric
    carbon dioxide concentration. The carbon dioxide released by
    combustion of fuel increases the flow of carbon from the rocks to the
    atmosphere. Deforestation increases the flow of carbon dioxide from
    the land surface to the atmosphere.

Whether or not these effects on the composition of the atmosphere are
important depends upon the *size* of the flows involved, by comparison with
the size of the reservoirs in the atmosphere and in the ocean. Some of these
quantities are shown in Table 5.4, but you should remember that global
quantities of this kind can only be estimated very approximately. Note that
the quantities transferred to and from the atmosphere do not balance,
because all the flows are not given in this table.

Q   Which flows *to* the atmosphere are missing from Table 5.4?

A   From marine and land organisms to the atmosphere through
    respiration and decay, and from volcanoes outgassing carbon dioxide
    to the atmosphere.

*Table 5.4   The quantity of carbon in the major reservoirs and the rate of transfer in
some of the flows between them*

Reservoirs	Carbon/10^9 tonnes
Atmosphere	700
Oceans:	
dissolved carbon dioxide	37 000
dissolved residues from dead organisms	980
living marine organisms	3
marine sediments:	
calcium carbonate	50 000 000
other residues from dead organisms	10 000 000
On the land surface:	
living land organisms	600
soil humus	3000
Fossil fuels	5000

Flows	Carbon/10^9 tonnes per year
Atmosphere to marine organisms	45
Atmosphere to land organisms	70
Oceans to marine sediments	1–10
Fossil fuels to atmosphere	5

Source: Trabalka, J. K. & Reichle, D. E. (eds) (1986) *The Changing Carbon Cycle: a global
analysis*, Springer, New York, Figure 20.2, p. 408.

An important flow of carbon dioxide from the atmosphere to the oceans is also omitted from Table 5.4. Carbon dioxide dissolves in rainfall, which then finds its way to the sea. The carbon in the sea is subdivided into that which is contained in dissolved carbon dioxide (as in very weak soda water), that contained in the residues of dead organisms (like the liquid in a very thin fish soup!), and that which is contained in living marine organisms. Note that the first of these quantities is by far the largest of the three. The carbon in marine sediments far outweighs all the carbon, dissolved or floating, in the body of the sea itself. This is because it has been accumulating there for millions of years.

Q   What proportion of the atmospheric carbon reservoir does the annual input of carbon dioxide from the combustion of fossil fuel represent?

A   5/700 = 0.007, or 7/10 of 1%.

This may not seem a lot, but it is actually an appreciable input. If the atmospheric reservoir absorbed all of this it would increase by 7% in just ten years.

Since 1957 measurements of atmospheric carbon dioxide concentration have been made on the peak of the mountainous island of Mauna Loa, Hawaii, in the middle of the Pacific Ocean. These records, which are taken at an altitude of nearly 3400 m, and are consistent with measurements made elsewhere, show that the average annual concentration of carbon dioxide has been increasing (Figure 5.14), and that the average annual concentration rose from 315 parts per million (p.p.m.) in 1957 to about 350 p.p.m. in 1987.

Q   What percentage increase does this represent?

A   The absolute increase was 35 p.p.m., which is

$$\frac{35}{315} \times 100$$

that is, 11% in 30 years, or 3.7% a decade.

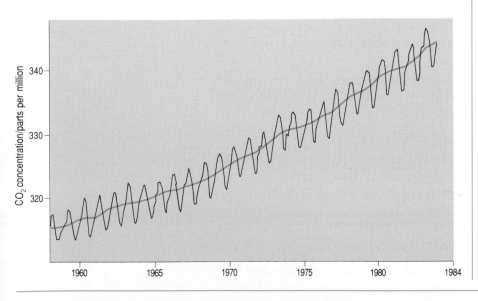

◀   Figure 5.14
The record of atmospheric carbon dioxide concentration from Mauna Loa Observatory, Hawaii.

This is about half the rate of increase that you calculated from the quantities in Table 5.4. What happened to the missing carbon dioxide? All the carbon dioxide which enters the atmosphere from the combustion of fossil fuel does not remain there. Some may be taken up again by faster plant growth, but most of the 'missing half' disappears. It is thought that it dissolves into the oceans, but what happens to it after that is not known.

We have already seen (Section 4) that there has been a rise in global mean temperature of 0.5 °C over the last hundred years. It is estimated that atmospheric carbon dioxide concentration was about 270 p.p.m. in the mid-nineteenth century and some scientists believe that the 30% increase since then is largely responsible for this rise in temperature, due to the greenhouse effect.

Q   As well as showing a steady overall increase in carbon dioxide concentration, Figure 5.14 shows another very obvious pattern. What is this?

A   There are very distinct, regular annual fluctuations in carbon dioxide concentration, with their peaks falling in the northern hemisphere spring of each year.

These regular cycles of carbon dioxide concentration are very much smaller in the southern hemisphere and have a larger amplitude (the distance between a peak and a trough in a cycle) the further north measurements are taken. Mauna Loa is at latitude 20° N. These cycles appear to be caused by seasonal changes in the growth and photosynthetic activity of land plants. There is much more land in the northern hemisphere than in the southern hemisphere, hence the smaller amplitude of fluctuations in carbon dioxide concentration measured in the latter.

The relationship between plants and the atmosphere is a reciprocal one. Trees in the forests of the northern temperate zone show an annual cycle of photosynthetic activity which is measurably correlated with (and seems to cause) seasonal fluctuations in atmospheric carbon dioxide concentration in Alaska.

It has been suggested that the rise in atmospheric carbon dioxide concentration might be reversed by planting trees on a huge scale.

Q   What assumption must be made about the rate of **decomposition** of wood, if planting trees is to have any impact on atmospheric levels of carbon dioxide?

A   Because decomposition releases carbon dioxide, we must assume that wood decomposes slowly if carbon dioxide fixed by trees is to remain locked up on land, and not be rapidly released back into the atmosphere.

Q   What calculation should you make to determine the rate of loss of carbon dioxide from wood?

A   You need to calculate the residence time of carbon in the living and dead tissue of forests.

In this section we have seen how the atmosphere protects living organisms from excess ultra-violet, and of course provides carbon dioxide for photosynthesis and oxygen for respiration. We have also seen that the relationship between life and the atmosphere is a reciprocal one, and that atmospheric composition has been changed by the cumulative effect of organisms' activities over geological time. The atmosphere also provides a vital blanket over the planet, raising and stabilising its temperature.

6.3   Carbon dioxide and ice ages

Neither the temperature of the Earth nor its atmospheric composition is stable, since we know that both have changed significantly during geological time. An historical record of temperature and atmospheric composition can be found in polar ice in which samples of the Earth's atmosphere have been trapped in layer upon frozen layer. A core of ice over 2 km deep, representing the last 160 000 years of snow accumulation, was extracted from the ice cap at Vostock by Soviet Antarctic scientists. This core covered a period stretching back from the present, through all of the last glaciation and the interglacial before it, and into the end of the penultimate glaciation.

The air trapped in the **Vostock ice core** was carefully analysed for its carbon dioxide content in samples taken from different depths, and dates were calculated for each depth from the estimated rate of ice accumulation. A technique based upon an established correlation between temperature and the ratio of two forms of hydrogen in frozen water was used to infer air temperatures from the same samples.

The results from the Vostock ice core (Figure 5.15) show an impressively close correlation between atmospheric temperature and carbon dioxide concentration between 160 000 years ago and the present. They confirm results obtained from shallower cores, which show that atmospheric carbon dioxide concentration increased from about 190 p.p.m. to around 280 p.p.m. at the end of both of the last two glaciations. Perhaps even more significantly, minor changes in air temperature during the last interglacial are also paralleled in the record of atmospheric carbon dioxide concentration.

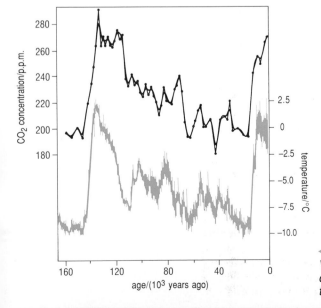

◄ Figure 5.15   The atmospheric temperature at Vostock in the Antarctic and the atmospheric carbon dioxide concentration from 160 000 years ago until the present, as determined from analysis of an ice core.

There is no unequivocal evidence from the Vostock ice core to tell us what the climatic mechanism is that links atmospheric carbon dioxide concentration and temperature. One theory is that the greenhouse properties of carbon dioxide are involved. The authors of the study believe that their evidence supports the theory advanced by a Serbian physicist, M. Milankovitch, in 1941: that the onset of glaciation is caused by certain periodic variations in the Earth's orbit around the sun: its ellipticity, the tilt of the Earth's axis and the time of year of the equinoxes. Each of these has its own cycle, which can be detected by mathematical analysis of temperature change, and the carbon dioxide data from the Vostock core.

Regardless of what events may trigger glacial periods, the Vostock core shows that there is a link between climatic changes and the carbon cycle. One scenario, amongst several possible theories, is that changes in ocean currents and winds caused ocean upwellings to become stronger at the height of glacial periods, and that this increased marine productivity.

Q   Why should an increase in ocean upwelling increase marine productivity?

A   Because upwellings bring nutrients to the surface of the sea where phytoplankton can use them.

Q   What effect would increased phytoplankton productivity have upon atmospheric carbon dioxide concentration, and why?

A   Greater phytoplankton productivity would cause greater fixation of carbon dioxide, reducing its concentration in the atmosphere if (and only if) this was then deposited in the sediments as skeletons or undecomposed material.

Note that this theory suggests that changes in atmospheric carbon dioxide concentration are caused, indirectly, by changes in global temperature, and not the other way round as a greenhouse theory would predict.

As we have already seen, relatively minor changes in temperature, from whatever their cause, can be amplified by their effect on the increase or decrease in the proportion of the globe's surface that is covered by polar ice.

Activity 7

What kind of evidence exists to suggest that atmospheric carbon dioxide concentration and global temperature were linked during past periods of climatic change?

6.4   Summary

The flows of carbon between rocks and sediments, the oceans, the atmosphere and living things are traced in the carbon cycle. The inputs and outputs of carbon dioxide to the atmosphere are particularly important because of the role of carbon dioxide as a greenhouse gas. The atmospheric concentration of carbon dioxide has been rising since before the 1950s.

7   Conclusion and summary

In this chapter we have sped through Earth history and skated over the planet's surface, dipped into the oceans and soared into the stratosphere. If you feel a little breathless after this trip around the world in fewer than eight hours, much less eighty days, then don't be worried. Much of what you have learned will be useful in later chapters, and you will be able to come back here to refresh your memory. In the meantime, the essential points to remember are these:

The Earth has an ancient history that can only be meaningfully recounted on a geological timescale that far exceeds our ordinary experience or conception. Despite this, it is possible to find out a great deal about past Earth environments, and such investigations show that the appearance of life on the planet, some 3.8×10^9 years ago, initiated great changes. The accumulated results of life's activities are to be found in the composition of rocks and the atmosphere. The carbon cycle in particular shows how the Earth's geological features, the oceans and the atmosphere are all linked by the activities of living organisms. Living things, being an integral part of the Earth system, are also very sensitive to the conditions of their environment.

It is to this relationship between organisms and their environment that we turn our attention in the next two chapters.

Answers to Activities

Activity 1

Q1   The three main types of rock are igneous, formed in the molten state; sedimentary, formed by deposition from wind or water; and metamorphic, transformed by heat and pressure.

Q2   **D.** Only living organisms are capable of reproduction.

Q3   Oxygen, and energy-supplying compounds of carbon.

Q4   Carbon dioxide and water.

Q5   Carbon dioxide and water.

Q6   Oxygen and compounds of carbon.

Q7   Living organisms are responsible for the high concentration of oxygen and the low concentration of carbon dioxide in the atmosphere. The presence of stromatolites and red beds in Precambrian rocks are evidence of this.

Q8   The carbon dioxide that plants fix by photosynthesis from the atmosphere and the oxygen they release in the process are both consumed when their tissues are burnt, or respired, by heterotrophes.

Q9   These plants consume carbon dioxide through photosynthesis and, in a separate process, in the formation of their hard shells.

Activity 2

Q1   Most water vapour is near the Earth's surface, in the troposphere. The ozone layer is in the stratosphere.

Q2   The troposphere is warmed by heat from the Earth's surface, so it gets cooler as the surface recedes. The stratosphere is warmed by the absorption of sunlight in gases, particularly ozone, so it gets hotter with altitude where least light has been filtered out.

Q3   Carbon dioxide, methane, water vapour, ozone and nitrous oxide are all naturally occurring greenhouse gases. They are called by this name because they insulate the surface of the Earth, allowing sunlight through but being relatively opaque to radiated heat.

Q4   Ozone in the ozone layer is formed by the action of ultra-violet in sunlight upon oxygen. It filters out some of the ultra-violet from sunlight, protecting life from its harmful effects.

Q5   The albedo of a surface is the proportion of light which it reflects. It is important to climate because there are big differences between the albedo of different surfaces on Earth, and so the relative extent of surfaces with high albedo (e.g. ice) relative to those with low albedo (e.g. vegetation) determines how much heat the Earth's surface absorbs.

Q6   At temperatures low enough to allow the ice cap to spread, its high albedo will cause further cooling. At temperatures high enough to permit the ice cap to melt, surfaces with lower albedo will be uncovered, and cause further warming. This is an example of a process called *positive* feedback.

Q7   As the ocean surface warms, there is greater evaporation, clouds form, and these cool the surface. As the surface cools, evaporation and cloud formation decrease, allowing the surface to warm up again. This is an example of a process called *negative* feedback.

Q8   Clouds will be most frequent in low pressure areas. As a result more solar radiation will be reflected back to space or absorbed in clouds and hence fail to reach the surface. This keeps equatorial areas cooler and means that solar radiation received at the surface is greatest in the suptropical high pressure areas, especially in summer. The cooling effect on mid-latitude areas is outweighed by heat transferred in winds and ocean currents.

Activity 4

Q1   An increase in global temperature can cause a rise in sea-level by the thermal expansion of sea water, and by the melting of ice on land.

Q2   Because of its curvature, the Earth receives sunlight at higher intensity at the equator than at the poles. This creates a gigantic convection cell with belts of high pressure at about 30° N and 30° S, especially over the ocean basins. Winds blowing out of these high pressure areas are deflected by the Earth's rotation causing them to circulate clockwise in the northern hemisphere and anticlockwise in the southern hemisphere. These winds then power surface ocean currents with the same general pattern.

Q3   Marine life depends upon the nutrients which phytoplankton need to grow. These nutrients tend to be removed from surface waters where phytoplankton grow when dead organic matter sinks into deep water. Where a thermocline develops, mixing between surface and deep waters traps nutrients in the sediment. Upwellings of cold water from the deep bring nutrients to the surface where they are accessible to phytoplankton.

Activity 5

Q1   One million is 10^6 and 1.5×10^3 is the same as 0.0015×10^6, so
$$a+b = 1.0015 \times 10^6$$

Q2   One hundred is 1×10^2, and 1.5×10^3 is the same as 15×10^2, so
$$a-d = 12 \times 10^2$$

Q3   To multiply two numbers using powers of ten, you add the powers and multiply the left-hand sides of the two numbers, so
$$b \times c = (1 \times 4.8) \times 10^{6+9} = 4.8 \times 10^{15}$$

Q4   To divide two numbers using powers of ten, you subtract the powers and divide the left-hand sides of the two numbers, so

$$\frac{a}{d} = \frac{1.5}{3} \, 10^{3-2} = 0.5 \times 10^1$$

Remember that 10^1 is just 10, so $0.5 \times 10^1 = 5$.

Q5   The residence time we calculated in Section 5.2 was 11.2 days. **E** is the answer nearest to this.

Q6   Trees increase the flow of water vapour from land to atmosphere through transpiration, and they reduce the loss of water from soil to rivers by trapping it with their roots.

Activity 6

Q1   Carbon is exchanged with the atmosphere in the form of carbon dioxide. Carbon dioxide is produced by the *respiration* of organisms on land (d) and in the ocean (a), and by the *combustion* of fossil fuels on land (c). Volcanoes also emit carbon dioxide. Carbon dioxide enters the ocean from the atmosphere by *dissolving* in sea water (b) and is transferred from the air to the land surface by *photosynthesis* (e). The carbon dioxide fixed by phytoplankton and algae in the sea comes from the carbon dioxide dissolved in sea water. Carbon dioxide produced by the respiration of marine organisms also dissolves in the water, rather than entering the atmosphere directly. Finally, carbon becomes incorporated in the rocks mainly by *deposition of calcium carbonate* and in carbon residues from dead organisms at the bottom of the ocean (f).

Activity 7

Samples of trapped air from the Vostock ice core show that atmospheric carbon dioxide concentration and global temperature were linked during recent glacial periods. The causes behind the link are unclear.

1   Introduction

1.1   Aims

This chapter and the next are concerned with the **biosphere** – that is, the living component of the globe. It should be clear by now that living things have played a central part in the history and development of the Earth as an environment, and that this development is crucial to the very existence of life on the planet. For example, the living inhabitants of the biosphere have influenced the composition of the atmosphere by lowering the concentration of carbon dioxide, by raising the concentration of oxygen, and in so doing by creating the ozone layer. Limestone rocks are composed of the skeletons of marine organisms, coal measures are the remains of ancient land plants, and the water cycle and local climates are strongly influenced by life on land, particularly the evapotranspiration of trees.

Now that we have studied Earth as an environment for life, it is timely to look at the inhabitants of the biosphere themselves, among whom we should of course number our own species. A sceptic might demand why we bother with this, and suggest that the only important question should be: 'What use are other creatures to us, and which ones can we do without?' An increasing number of people regard such an attitude as a kind of biological chauvinism on the part of humans, and invoke aesthetic or moral reasons for preserving all other species (as considered in Chapter 3). However, the very survival of biological diversity (commonly abbreviated to biodiversity) may depend upon being able to show the sceptics that it is in their enlightened self-interest to maintain the diversity of living things. We are presently in ignorance of one of the most basic facts: we don't know how many species there actually are. The pragmatic answer to the sceptic is that we cannot tell what use everything might have until we have an inventory and can understand what role different organisms play in the communities in which they live.

Chapter 5 was a whirlwind global historical tour, but to get to grips with Earth's inhabitants and their activities properly, it is now necessary to change gear and to introduce some detail. Plants provide not only the fuel tank of the biosphere but also, on land, provide the physical framework that shelters animals. So, we will begin by looking at the relationship between different regional climates and the characteristic kinds of vegetation, called **biomes**, that develop in them.

Next, in a kind of biological inventory, we will attempt to sketch what kinds of organisms share this planet with us, where they are, and what they do. This includes seeing how animals and plants are classified, looking at estimates of how many species there are, and reviewing the modern threat of extinction of species, a major part of which is due to the destruction by human activities of the places where wild species live. The local environment where an individual animal or a plant lives is referred to as its **habitat**. No animal or plant lives an existence independent of other

As you read this chapter, look out for answers to the following key questions.
- What are the world's major biomes?
- What are the difficulties in finding out what species exist and which of them are in danger?
- How are organisms related to habitats in oakwoods, heathland and the rocky seashore?

animals and plants, so we must pay attention to the larger communities of which all living organisms form a part. In the final section of this chapter three familiar British habitats are examined, and you will discover that their character and preservation depend upon a proper understanding of their ecology, and that human activities play an important role.

1.2   Ecology

The science of **ecology** is the study of the relationships of organisms with each other and with their environment. Humans should, of course, be included. Ecology is potentially a huge subject. Indeed, professional ecologists tend to feel humbled as well as fascinated by the size and complexity of ecological systems.

To try to bring order to this complexity ecologists use a variety of different ways of looking at environments. For instance, habitat is an organism-centred description of the environment: primroses are found in certain kinds of woodland habitat, mallard ducks are found in certain aquatic habitats, and so on. When we go into the ecological relationships between living things in more detail in Chapter 7, we will find that there is another way of looking at organisms, which stresses their interrelationships and interdependence as a part of an **ecosystem**. Habitat and ecosystem are alternative, and to some extent complementary, ways of looking at ecological relationships. The concept of *habitat* is useful for describing *where* certain animals and plants are to be found, and *with whom* they live. The concept of *ecosystem* is useful in helping to explain *how* these relationships function.

1.3   Stocktaking our dwindling living resources

No habitat has been studied in sufficient detail for us to be able to enumerate every organism that is to be found there. In an ideal world one might want to investigate all available habitats in ecological detail but the fact is that because of human activities, habitats are disappearing faster than they can be studied, and simple stocktaking of threatened habitats is now an urgent task. It isn't rational to use a treasure-chest for firewood until you've checked inside for treasure, yet this is what is happening to the Earth's largely uncatalogued biological treasure-houses. Foremost among these are the tropical rainforests.

There is plenty of evidence that tropical rainforests contain a genuine treasure trove of potentially useful and economically important species. The majority of people in the developing world rely upon traditional medicine, which depends heavily on the use of plant extracts. Many medicines in the developed world also come from plants. It has been estimated that about half the pharmaceuticals used in the United States contain active constituents derived from tropical plants. A medicine useful against childhood leukaemia comes from the Madagascar periwinkle, which is a tropical forest plant. Although there are 119 plant-derived pharmaceuticals in use throughout the world, these originate from fewer than 90 species of plant, so the potential for finding new medicines is huge.

Wild species with novel economic and medical uses are not entirely confined to tropical rainforest. Jojoba oil from a North America desert shrub has been introduced as a replacement for sperm whale oil, and finds increasing use in cosmetics. At higher temperatures *Tilapia* fish native to Lake Malawi are more efficient than carp at converting food into flesh, and

have proved profitable in fish farms. Apart from humans, armadillos are the only animal that can contract leprosy, and they have proved useful in research into the production of a vaccine. Many wild species like these, but whose uses are still unknown, are in danger of extinction. Four-fifths of the tropical forests of Madagascar, which yielded the Madagascar periwinkle, are now gone.

2   Climate and vegetation

2.1   Climate

Plants and animals living on land are sensitive to the temperature and moisture in their environment. These two climatic factors, and their seasonal variation, are the major influence on the kinds of vegetation found in different parts of the globe. The relationship of vegetation to climate is therefore a logical place to start an inventory of the biosphere.

Both rainfall and temperature change in a systematic manner with latitude and with altitude, creating recognisable geographical patterns in the vegetation. Figure 6.1 illustrates the major regional types of vegetation, or biomes, for an 'average continent' created by generalising the patterns found on the actual continents. The shape of the average continent has a familiar appearance because, like Africa, South America and India, it tapers from north to south, thus representing the distribution of land area on the real globe at the present day. The areas of each biome depicted on the average continent show the *potential* extent of forests, savannah and so on, with limits set by climatic conditions. The biomes are the nearest thing to the 'natural vegetation' of a region. Figure 6.1 does not show the *actual* extent of the biomes because agriculture and other human activities have greatly affected this.

Q   What is the major pattern shown by the vegetation types in Figure 6.1?

A   At least in the northern hemisphere, the more extensive vegetation types occur in bands that approximately follow lines of latitude.

In the southern hemisphere, the average continent tapers, so its coastal climate and vegetation are more strongly affected by the influence of the ocean and by mountain ranges.

Q   Which major vegetation types of the northern hemisphere are entirely missing from the southern hemisphere?

A   Most of those from the far north: tundra, boreal forest and steppe.

The absence of these vegetation types from the southern hemisphere is simply due to the absence of large land masses at the appropriate latitudes in the southern half of the globe. Remember that in Chapter 5 we saw that atmospheric carbon dioxide concentration fluctuates through the year (Figure 5.14), and that these fluctuations are stronger in the northern hemisphere than in the south because there is extensive boreal forest only in the north.

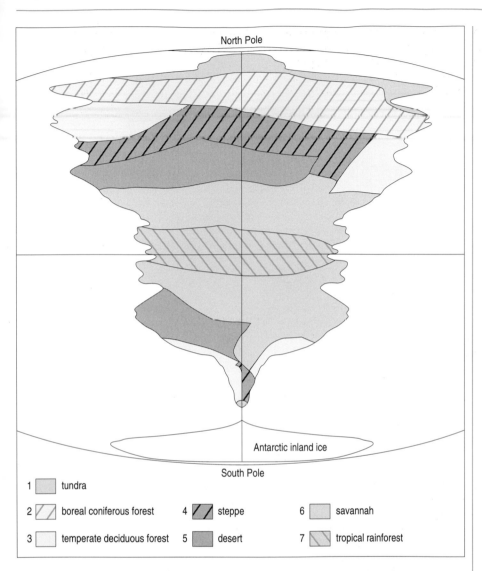

North Pole

Antarctic inland ice

South Pole

1		tundra
2	///	boreal coniferous forest
3		temperate deciduous forest
4	///	steppe
5		desert
6		savannah
7		tropical rainforest

▲   *Figure 6.1   The major types of vegetation on an 'average continent', representing their potential geographical distribution on the earth as a whole.*

2.2   The world's biomes

The major vegetation types of the world are referred to as biomes.
Examples of these are illustrated in Plates 4 to 10. The composition of each
biome can be described in terms of its physical structure: for example,
whether it contains trees or not, whether these are scattered or create a
dense forest, and so on. The prairie grasslands of North America and the
grassland steppe of central Asia are separated by thousands of miles, but
both are recognisably representatives of the grassland biome. This does not
mean, however, that the same grass species are to be found in prairie and
steppe. Even within the same biome there is great biological diversity
between different geographical areas, because of the evolution of species
that has occurred in the great time which has elapsed since the origin of
most major groups of animals and plants.

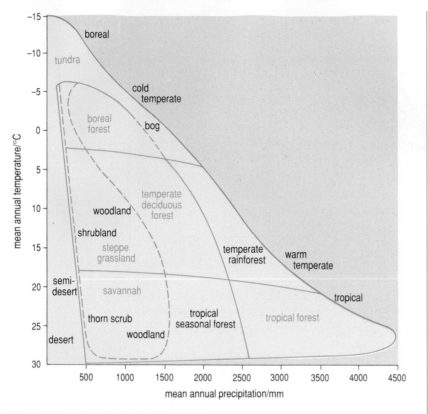

▲   Figure 6.2   Distribution of the major biomes in relation to average
annual precipitation and average annual temperature.

A combination of two principal enviromental variables determines this
structure: annual precipitation and annual temperature. The relationship of
biomes to averages of these two variables is shown in Figure 6.2. Notice
that temperature is on the vertical axis, and that the scale runs from +30 °C
at the bottom to –15 °C at the top. This is an unusual way of plotting a
variable (scales normally increase as you go *up* the axis), but it happens to
be more convenient to plot this diagram in this way. Study this figure and
then answer the questions below.

Q   In what kinds of conditions are trees absent?

A   Principally, when average annual rainfall is less than about 500 mm in
    hot climates, or when mean annual temperatures fall below about
    –5 °C. Notice that trees do occur where there is less than 500 mm
    rainfall in some colder climates, for instance in the boreal forest.

Q   If temperatures remained about the same, but average annual rainfall
    increased significantly, how would you expect the vegetation of
    prairies and tropical savannahs to change?

A   These would probably be invaded by trees and become temperate
    deciduous forest and tropical seasonal forest, respectively.

Q   If average annual temperatures increased by a few degrees, what
    would you expect to happen to areas at the edge of the tundra biome?

A   They would probably be invaded by trees and become boreal forest.
Some ecologists have suggested that this a real possibility if the
greenhouse effect causes the global climate to warm up.

Figures 6.1 and 6.2 are of course greatly generalised. In a place where the
average temperature is high, a single night of unusual frost could alter the
vegetation locally. Altitude is also an important influence on local climate
and vegetation.

Q   You have already seen in Chapter 5 how temperature changes with
altitude. Using this knowledge and Figure 6.2, how would you expect
the vegetation on a tall mountain in the tropics to change as you climb
from low altitude to its peak?

A   Temperature generally falls with altitude. As you ascend, you would
expect to pass from tropical forest at the foot of the mountain,
through a temperate-like deciduous forest, to a boreal-like coniferous
forest, before emerging into a treeless zone with the appearance of
tundra.

The upper limit of altitude for tree growth is called the tree-line. The
tree-line occurs at lower altitudes on mountains in the temperate zone than
in the tropics, because temperatures are lower in the former area, even at
sea-level.

2.3   Summary

Average annual rainfall and temperature vary in a systematic manner with
latitude and altitude, producing characteristic regional types of vegetation
called biomes. Going from the equator northwards the principal biomes are
tropical rainforest, tropical seasonal forest, savannah, desert, steppe,
temperate deciduous forest, boreal coniferous forest and tundra.

3   An inventory of the biosphere

3.1   Taxonomic classification

Any inventory requires a systematic scheme for describing the items we
are going to put on our list. The science of classifying organisms, putting
names to them and working out relationships between them is called
taxonomy.

All organisms, living and extinct, are thought to have evolved from a
common ancestor. This common descent imposes a natural order upon the
classification of living things based upon their degree of relatedness. This
can be represented as a tree with a trunk and branches. At the tips of the
branches of the taxonomic tree are the units that are called **species**. The
common ancestor of all life is the root of the tree and from this sprout the
major groups of the living world, called **kingdoms**. There are alternative

views as to how exactly the kingdoms should be divided, and we shall use a classification that recognises four of them: the **fungi**, the **plants**, the **animals**, and the **prokaryotes** (**bacteria** and blue-green bacteria).

Following the system devised by the eighteenth-century Swedish biologist Carl Linnaeus, each species is given a scientific name in Latin, made up of two parts. The first part of the name is that of the **genus** to which a species belongs (e.g. *Homo*), and is shared with closely related species. The second part of the name is that of the species itself and is unique. Humans are classified as a single species, *Homo sapiens*. We have no living relatives thought similar enough to be placed in the same genus.

Scientific names are preferred to common names because they are internationally recognised and are less confusing. For example the plant known by the scientific name of *Chenopodium bonus-henricus* is called 'good King Henry' in some parts of Britain and 'fat hen' in others. In some places fat hen is also the name used for another species, *Chenopodium album*.

Taxonomists use a hierarchical classification system to describe how species are arranged with respect to each other and how they fit into the branching pattern of the tree. Between genus and kingdom, taxonomists recognise several other levels of classification. To simplify this somewhat, genera are grouped into families, families into orders, orders into classes, and classes into phyla. These are no more than artificial ways of dividing up species into different sorts with certain features in common. In the animal kingdom, there are many more phyla than in the plant kingdom. The largest phylum of animals is called the arthropods, which includes the insects as a class. The beetles are within an order of the insect class.

Biologically, the species is the only taxonomic unit that nature recognises. It is argued that the category 'species' is different from all the others, and that 'nature recognises it' because it is the largest group that is held together, and indeed defined, by a biological process: namely, mating among its members. So, a species is a group of organisms that are able to interbreed in nature. In practice, taxonomists, who usually work with dead specimens, can never apply this criterion to discover whether two individuals belong to the same species. Instead taxonomists use physical characteristics such as number of hairs on the head of a fly to determine the species. Unfortunately, no two members of the same species are identical, unless they are twins produced from the division of a single fertilised egg. The variation in appearance between individuals of the same species creates difficulties for taxonomists. It is not unknown for the males and females of a species in which the two sexes look quite different from each other to be misclassified as belonging to separate species, until someone discovers them mating in the wild.

It is especially difficult to apply the concept of a biological species to bacteria, which multiply very rapidly by asexual means. Bacteria *are* given species names (for example, the common gut bacterium in our stomachs is *Escherichia coli*), but there is very great diversity of behaviour between different strains of the same bacterial 'species'. Consequently the classification originally devised for animals and plants, and which fits these organisms reasonably well, is less satisfactory for bacteria.

The taxonomic tree shown in Figure 6.3 is constructed on the basis of the similarities and differences between living species, but since we have good grounds to believe that all species are ultimately related through a common ancestor, such trees can also be seen as mapping out the pathway of evolution. The taxonomic tree can also be viewed as an evolutionary one. We, and all other living species, sit perched on the tips of the branches of the evolutionary tree. What does the evolutionary tree tell us about living species?

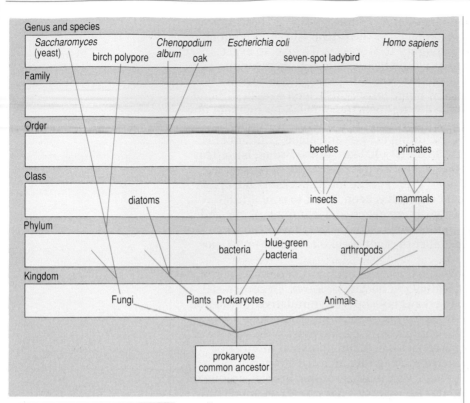

▲   *Figure 6.3   The four kingdoms, and some living animal and plant groups mentioned in this chapter, arranged in a taxonomic tree. Different levels in the hierarchical taxonomic classification are indicated by horizontal boxes. By convention the scientific names of species are written in italic script.*

First, note that we must be cautious in interpreting what the evolutionary tree means. We share our position at the tips of the tree's branches with all other living species. There is nothing exalted in this position: the simplest worm shares it with us, as much as do chimpanzees and sunflowers. There is no superior or inferior in evolution – only survival and extinction. We must also remember that millions of species, many major groups such as the dinosaurs and even whole phyla have gone extinct during evolutionary time and that these dead branches of the evolutionary tree are invisible today. What we are looking at in Figure 6.3 is a pattern inferred, save for the evidence of fossils, largely in ignorance of what has *actually* gone before.

What the evolutionary tree does tell us quite unequivocally is that new species have been continually produced in the history of life, and that evolution generates diversity. Hence we are led to ask, how many species survive today and what are their risks of extinction?

3.2   How many species are there?

The most straightforward answer to this question is that no one knows. About 1.82 million species have been given a scientific name using the system established by Linnaeus in 1753, but these are a highly selective and very incomplete sample of the whole. The best we can do is to try to estimate how the sample is biased, and to what degree it is incomplete.

Plants are the best known kingdom and some 250 000 of these have been named. The flora (i.e. list of plants) of the British Isles is as completely described as any and contains roughly 1600 species. To give an exact number even for the British flora would be to invite arguments with botanists about the number of species in certain plant groups that are taxonomically difficult to classify! Some taxonomists tend to split difficult groups into many species, while the 'lumpers' favour classifying them into only a few. Even if we conceded to the most nit-picking of taxonomists, the total number of plant species, excluding mosses and algae, could not be pushed above 2000 to be found in the 315 000 km^2 of the British Isles. This total includes around 70 species of tree. To put this in global perspective, a single hectare (100 hectares (ha) = 1 km^2) of the most species-rich tropical rainforest in the upper Amazon region has been found to contain 283 species of tree alone. Tropical forests are also rich in climbing plants and epiphytes (plants which live in the branches of trees) whose numbers would surely push the total number of plant species to well over 300 per hectare.

In forests as rich as those of the upper Amazon most species are rare and have densities of one individual per hectare or lower. Of course 300 new species are not found in every hectare, but the cumulative number of species encountered increases rapidly with each additional hectare that is sampled. The rate at which this cumulative total increases with the size of the sample area provides one means of calculating the total number of species, named and unnamed, in a much larger area. This rate is determined by the extent of species' distribution ranges. Species unique to particular areas are called endemics and push up the total: regions with large numbers of species commonly have high rates of endemism.

In Britain there are very few endemic plants (all but 16 occur somewhere else) and the number of plant species is known, probably to within 99.9% accuracy. The insect fauna of the British Isles is much larger (about 23 500 species) of which perhaps 90% have been named. This level of knowledge is highly atypical of most parts of the world, particularly in the tropics, where there are more species and much less is known about them.

An important discovery that has been made in samples of well-described flora and fauna is that the vast majority of species are rare. In any one place, only a small minority of species can be called abundant. This is a general rule which appears to apply with extraordinary reliability to all samples of living things that have been collected. It is useful to know that rarity is the rule rather than the exception for the average species. For one thing, most ecological studies are conducted on common species, and so we must remember that these may be unrepresentative of the species most at risk. One compensation for this is that what is rare in one place may often be common in another. Also, it is not unknown for populations of rare species to suddenly become common, and even to reach pest status when conditions change. This happened to the pine beauty moth which was a rarity in Britain until it began to attack the new plantations of lodgepole pine in Scotland.

The difficulty in counting numbers of species in habitats such as tropical forest is the sheer size of the task. Identification of species that have already been formally named is difficult enough, but large numbers of undescribed species are usually encountered, particularly among insects and spiders. These require the attention of a specialist in the taxonomic group in question, and usually there are very few of these. E. O. Wilson estimated in 1985 that there were only 1500 taxonomists world wide who were qualified to deal with the diversity of tropical organisms. Even for groups like the ants and termites, which are of major ecological and economic importance in the tropics, Wilson knew of only five people in the entire world who devoted themselves full time to their description and classification.

In order to estimate the total number of species on Earth for any particular taxonomic group, we have to extrapolate from small samples which have been thoroughly surveyed and enumerated to the entire area occupied by similar habitat. On the basis of such methods, which only amount to using an educated guess, it has been estimated that there are 50 000 species of tropical trees as a whole.

The animal kingdom represents more than 70% of all named species, and of these 41 000 are vertebrates (fish, birds, reptiles and mammals) and 751 000 are insects (Figure 6.4). Three things *can* be reliably deduced about the composition of the total of all living species from the figures for the numbers of described species:

1   There are many more animals than plants;

2   Most animals are insects;

3   Most insects are beetles.

Bacteria and prokaryotes are generally left out of such calculations, because they are so difficult to count and to classify, but we must not forget them altogether, for they are ecologically vital, as we shall see in Chapter 7.

Extrapolating from the number of described species has produced estimates for the total number of living species ranging from 3 million to 30 million. The higher estimate was produced after an entomologist studying insects in tropical forest sprayed insecticide to 'knock down' insects from the canopies of trees. The tree canopy is an inaccessible and mostly unexplored layer of the tropical forest. This new study yielded a wealth of new finds. Over 1100 species of beetles alone were gathered from the

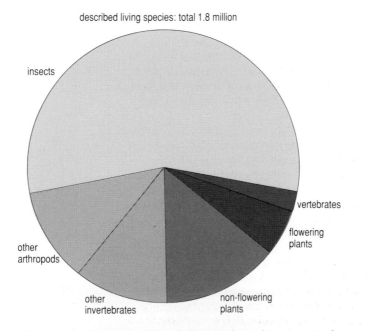

▲   Figure 6.4   The proportion of all named species accounted for by the major groups of animals and plants.

canopies of one tree species, *Luehea seemannii*, and it was thought that 160 of the beetles in the sample were able to live only on this species of tree, i.e. that they were endemic to this tree species. If all the estimated 50 000 tropical tree species each have this number of endemic beetles in their canopies, then this specialised beetle fauna alone must number 8 million. Regardless of the accuracy of this calculation, it is not difficult to see why, when a clergyman asked the biologist J. B. S. Haldane what he could deduce about the Creator from His creation, Haldane replied, 'An inordinate fondness for beetles'. About a quarter of all described species are beetles. No good scientific reason for the preponderance of this group is known.

Although numerous in species, beetles are outnumbered in *individuals* by other groups, and the tree canopy may not be the place to look for the bulk of the fauna in a tropical forest. One study which sampled the insects, spiders and mites of the soil as well as the tree canopy in a tropical forest in Indonesia, estimated that there were about twice as many of these animals in the soil as in the canopy. The commonest animals in the soil were springtails (tiny insects) and mites, which both outnumbered the individual beetles found there. Perhaps the preponderance of beetles among *described* species partly reflects taxonomist's fondness for sampling above ground rather than below.

3.3   *Where are these species?*

Given the inaccuracies inevitably involved in estimating how many species there are on Earth, it is perhaps more important to know where the most species are to be found.

For many different taxonomic groups there is a gradient in the number of species between the equator and the poles. Because the plant kingdom is the best catalogued, this trend is easiest to see for this group (Figure 6.5). Although other causes may be contributory, the geologically recent glaciation (which ended about 13 000 years ago) experienced in the temperate zone is possibly a significant cause of the pattern shown in Figure 6.5. Glaciation (see Chapter 1) severely reduced the flora and fauna of northern latitudes, so that most of the species present there today are recent recolonists. The process of recolonisation is so slow for some species, for example the chestnut in North America, that they are still advancing from their southern refuges from glaciated areas.

Tropical forests cover about 7% of the world's land surface, but harbour perhaps half of the world's species. Along with the great tree and insect diversity already mentioned, this biome also accounts for most of the world's primate species: our closest relatives. Just four tropical countries, Brazil, Madagascar, Zaïre and Indonesia, are home to 75% of these animals. The World Wide Fund for Nature identifies these four nations, plus Mexico and Colombia, as 'megadiversity' countries on the basis of their biological riches (Figure 6.6).

The designation of certain 'megadiversity' countries is intended to pinpoint the areas most in need of protection, but it is by no means a complete picture of the world's centres of biological diversity. Other parts of the world may be less diverse in their overall biological richness, but have exceptional numbers of species belonging to particular groups. For example, Lake Baikal in the Soviet Union is exceptionally rich in fish species, as are Lake Victoria and Lake Malawi in Africa, but Baikal has more endemic species than any other lake in the world. The most diverse fauna of freshwater molluscs (snails, mussels, etc.) occurs in the waters of central and eastern North America. The island continent of Australia contains a unique flora and fauna with a very high rate of endemism. The tree genus *Eucalyptus* is almost entirely endemic to Australia, though many of its 600 species have been imported into tropical and some warm

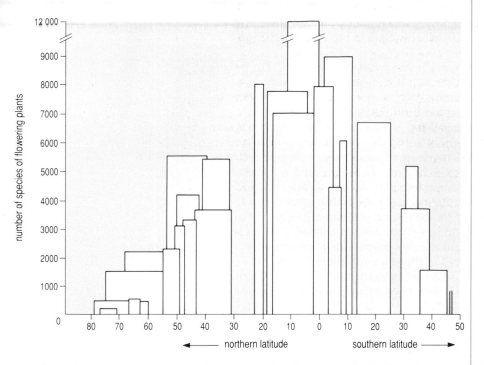

Figure 6.5   The number of species of flowering plants listed in the floras of the world. Each column represents a country spanning the latitude shown on the horizontal axis.

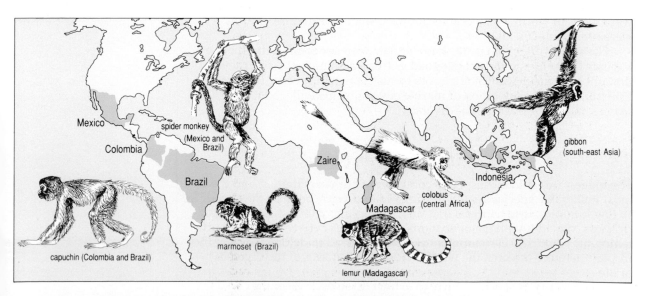

Figure 6.6   Megadiversity countries identified by the World Wide Fund for Nature.

temperate countries where they provide extremely fast growing timber. The Great Barrier Reef of Australia is also a centre of marine diversity.

It has been estimated that only about 20% of all living species are to be found in the sea, though these include representatives of nearly all the phyla – many of which are not to be found on land at all. Most of the world's known animal species are insects. Because this group is almost entirely absent from the sea, it should not be too surprising that estimates of the number of species there are low. However, the vast expanse of the ocean, and its immense depth in places, make it biologically the least known part of the planet.

It would be unwise to underestimate what the exploration of the ocean realm may yet reveal. As recently as 1983 a whole new phylum was described to accommodate a single species of wormlike creature new to science, which had been discovered living in ocean sediment. The megamouth shark *Megachasma pelagios* is five metres long, but is known from only two specimens caught in a decade and a half. A whole new habitat was discovered in 1977 on deep-sea ocean ridges where tectonic plates divide (Chapter 5), and which can only be explored with remotely controlled cameras or with submersible craft able to withstand immense pressure. Rich animal communities were found living around hydrothermal vents in the Pacific sea floor that disgorge sulphide in hot currents of water saturated with minerals. These vents are 2500 m below the surface, so deep that no light reaches them. The giant tubeworms, crabs, and other strange animals found in these communities live upon chemosynthetic bacteria that are able to use sulphide instead of light as a source of energy. More than 100 new species have been discovered living in these environments.

Even the exciting discovery of the hitherto unsuspected communities around hydrothermal vents has not produced as many new species as have been encountered in samples taken from other deep sea areas. In these the animal communities depend entirely upon organic matter which sinks to the deep ocean from the surface. At depths of 1500–2500 m off the coast of New Jersey, USA, samples were collected by dragging the sea bed with a sampler device that can be closed before bringing it to the surface: this contained 898 species, nearly half of them unnamed. Most of the species were kinds of worm, crustacean or mollusc, but a dozen different phyla were represented in all.

Coral reefs are much better known than deep sea communities, but as habitats they are as poorly catalogued as tropical rainforests, and they are thought by some marine biologists to be just as rich in species. It looks likely that future exploration of marine environments will lead to a reassessment of global species diversity.

3.4   *Which species are in danger, and why?*

The biggest threat to most species comes from human activities, and the most vulnerable species are the rarities and endemics, because they occur in so few localities. It so happens that the distribution of terrestrial species richness coincides with major economic imbalances between tropical and temperate countries. Development policies that lead to deforestation in the tropics therefore threaten the world's major repositories of the diversity of life.

The majority of species which go extinct do so because of the destruction of their habitat. This is not a process confined to developing

◄ *Figure 6.7*
The large blue butterfly and
its ant host.

countries. Since 1949 Britain has lost 95% of lowland meadows, 50–60% of lowland heath, and 30% of upland heath and grassland. More than 300 plant species are on the official danger list. The large blue butterfly (*Maculinea arion*) was a recent casualty of habitat loss in Britain (Figure 6.7). This rare species was pushed to extinction by the shrinkage of calcareous grasslands (i.e. on chalk and limestone) in southern England. The caterpillars of the large blue spend a crucial part of their development in ants' nests, which are lost when calcareous grasslands are ploughed.

Although it has become extinct in Britain, the large blue is still to be found on the continent of Europe. Britain has only a handful of endemic species and extinctions are therefore of more local than global significance. Most species could therefore theoretically be reintroduced, but usually only with a great deal of effort, by bringing in individuals from a population elsewhere. None the less extinctions are to be avoided because the loss of a local population generally means a loss of genetic diversity, which itself can endanger the future of a species. The situation in tropical habitats with their characteristically high rates of endemism is certainly the most serious. Local extinctions in the tropics are more likely to be terminal for whole species.

Because of the poor state of knowledge about species in tropical habitats, most extinctions go unrecorded. If there are over 1000 species of beetle in the canopy of a single tropical tree, many of them unnamed, who is to know if the last of a particular species was the victim of forest clearance? Quite certainly if the large blue had been a tropical butterfly and was as rare in that habitat as it was in Britain before its demise, the chances are it would have disappeared from the face of the Earth before any scientist even knew of its existence. For these reasons estimates of extinction rates can only be very approximate, and they vary widely depending upon the assumptions that form the basis of the calculation. Estimates of the number of species that will have become extinct between 1980 and the year 2000 due to tropical deforestation vary between 4% and 50% of the Earth's total.

Q If we assume the lower extinction rate of 4% and the lowest estimate of the Earth's total species (3 million), what is the lowest estimate of the number of species that will have been lost between 1980 and the year 2000?

A 4% of 3 000 000 = 120 000 species.

Q What kinds of animals or plants would you expect most of these extinctions to affect?

A Insects, such as beetles, because there are so many species of them in tropical forests and elsewhere.

Numerical estimates of this kind are extremely uncertain, but they are based upon the assumption, sadly sound at the time of writing, that there is an accelerating rate of habitat destruction in the tropical forests that shelter a disproportionate concentration of the Earth's terrestrial species.

Modern extinction rates can be put in historical perspective by comparing them with those recorded by palaeontologists in the fossil record. Measured on the geological timescale, extinctions in the record show that there have been a number of pulses of species loss, called mass extinctions, when whole groups of animals met their end at about the same time. The most well known of these events was 65 million years ago at the boundary between the Cretaceous and Tertiary Periods, when both marine and terrestrial species disappeared. About 50% of marine genera vanished.

The extinction of species in the fossil record can only be dated with crude accuracy. Species that seem to have disappeared simultaneously may have actually become extinct tens of millions of years apart. Some groups, such as the dinosaurs, were declining before the dawn of the Tertiary, which finally saw their end. The suddenness of mass extinctions can only be judged on the geological timescale that is appropriate to it. This makes a comparison with modern extinction rates very difficult, though some palaeontologists have attempted to make one. David Raup, for example, has estimated that over the whole of the last 600 million years species extinction rates averaged 9% per million years. By comparison, the estimated modern extinction rate of 4% over 20 years is very rapid indeed.

Over a geological timescale species extinctions are compensated by the evolution of new species, but what is happening today is too fast for evolution to restock the biosphere. Even if estimates of modern and fossil extinction rates are wildly inaccurate, it is obvious that the world is facing an unprecedented crisis of extinction.

Activity 1

In the discussion of recent extinctions which follows, you should think about the different kinds of human activities that endanger wildlife, and about what makes some species more vulnerable than others.

The extinction of vertebrate animals does not generally go unrecorded and tells a sorry tale (Table 6.1). Ninety-four species of bird are known to have become extinct since 1600. The majority of them, like the dodo of Mauritius and its neighbour the solitaire of Rodriguez, were inhabitants of oceanic islands where populations are especially vulnerable because of their small size and their long isolation from predators. Exceptions are the pink-headed duck and the passenger pigeon which were both continental species.

Neither were all the extinct species rarities before their extirpation. The passenger pigeon was once so numerous that its flocks are said to have darkened the sky as they flew overhead. The blue pike was a fish of great importance in Lake Erie, where it constituted more than a quarter of the total commercial catch between 1915 and 1959. The harelip sucker, which was a North American fish of freshwaters and which went extinct at the turn of the twentieth century, was reckoned to be the most abundant and valuable fish of its kind only twenty years before the last specimen was caught in 1893.

For every animal or plant which is known to have been driven extinct in the last 200 years, there are many hundreds that have been reduced to the brink of extinction and which now teeter on the edge. The International Union for the Conservation of Nature (IUCN) co-ordinates a system of 'red data books' that list threatened species on a group-by-group basis for each country. The data are incomplete, but a conservative estimate is that 25 000 species of plants are threatened.

Q   What proportion of all named plants is this?

A   There are reckoned to be 250 000 named plants. So, 25 000/250 000 = 10%.

Q   Why is the number of threatened plants such a large fraction of the whole?

A   Most species are rare anyway, making them vulnerable to extinction.

Table 6.1   Some vertebrate extinctions

Species	Former distribution	Last recorded	Probable cause of extinction
Birds			
dodo	Mauritius	c.1681	hunting
great auk	North Atlantic	1844	hunting by sailors
passenger pigeon	North America	1889	hunting and habitat destruction
pink-headed duck	Bengal	1936	hunting
solitaire	Rodriguez	c.1791	hunting
Mammals			
auroch	Europe	1627	hunting and habitat destruction
Caribbean monk seal	Caribbean	c.1960	hunting
Steller's sea cow	North Pacific	1768	hunting
Tasmanian wolf	Tasmania	1900s	hunted as a 'pest'
Fish			
harelip sucker	rivers of North America	1893	fishing and habitat destruction
blue pike	Great Lakes, North America	1970	overfishing, pollution, predation by introduced fish species
Reptile			
St Croix racer	St Croix, US Virgin Isles	1900s	

Lake Victoria and Lake Malawi in Africa and Lake Baikal in the Soviet Union are ancient lakes with an exceptional diversity of fish species – the product of millennia of evolution in these isolated aquatic habitats. All are centres of diversity of global importance but Lake Victoria is irreparably damaged and Lake Baikal is threatened. The endemic fish of the African lakes all belong to a group called the cichlids (pronounced 'sic-lids').

Lake Victoria, which has an area of 68 635 km² (much larger than Switzerland), once supported important fisheries which harvested plant-eating and detritus-feeding cichlid species (Figure 6.8a). In 1960 an alien predatory fish, the Nile perch *Lates niloticus*, was introduced into the lake in the mistaken belief that it would improve the fishery, because it grows to a large size. As was predicted by ecologists whose advice was ignored, the Nile perch had the reverse effect. Since its introduction, nearly all the commercially important indigenous fish have disappeared (Figure 6.8b), creating an economic disaster and an incalculable loss to science. Scientists concerned about this dual catastrophe have warned that it could be repeated in Lake Malawi if alien fish are introduced there.

The threat to Lake Baikal and its unique fauna of 2500 species is of a different kind, but every bit as critical. The lake is huge (over 700 km in length and reaching a depth of 1640 m) and is the largest single body of freshwater on Earth. Despite this size, and water of legendary clarity, the lake is threatened by large discharges of waste water from pulp mills on its shores.

The destruction, pollution or alteration of wildlife habitats threatens the species found in them wholesale. In the next section we will look at three examples of habitats found in the British Isles, to see in a little detail what species live there, and how human activities can alter and shape the environment of these animals and plants.

3.5   Summary

The science of classifying organisms is called taxonomy. Because all living species are derived from a common ancestor, their relationships to each other can be described as a tree, with species at the tips of the branches, and four kingdoms sprouting from the trunk. The four kingdoms are: plants, animals, fungi and prokaryotes. The species is the only natural unit of classification, and is defined as a group whose members may all potentially interbreed. Species are given scientific names in Latin, consisting of two parts: the genus and the species, e.g. *Homo sapiens*.

About 1.82 million species have been named, but estimates of the actual total on Earth vary between 3 million and 30 million. Of the described species, plants are the most thoroughly known and number about 250 000 named species. Among described animal species, the insects are by far the most numerous, and most of these are beetles.

Two generalisations hold for most groups of animals and plants:

1   Most species in any locality are rare and very few are abundant.
2   The tropics hold many more species than the temperate zones.

Coral reefs and tropical rainforests are both rich in species but both are poorly catalogued. Most named species are terrestrial, but exploration of the sea bed has found many new, undescribed species.

The biggest wholesale threat to the survival of species comes from the destruction of habitats by human activity. Destruction of tropical

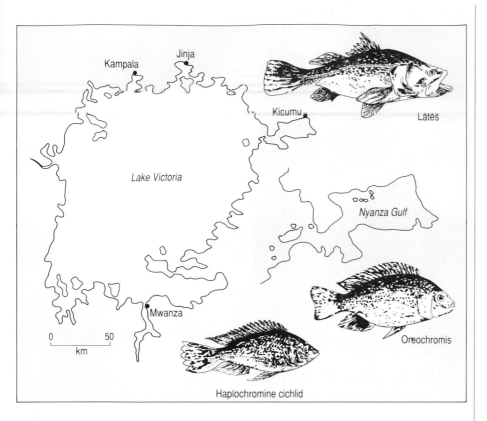

▲   *Figure 6.8(a)   Lake Victoria, Nile perch, and two endemic cichlid fish.*

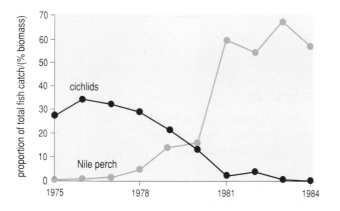

▲   *Figure 6.8(b)   Decline in the proportion of endemic cichlids and rise in the proportion of Nile perch in the fish catch in Kenyan waters following the introduction of the Nile perch to Lake Victoria.*

rainforests is the foremost example of this, but the process also occurs in the economically developed countries of the temperate zone, including Britain. Although the extinction of species has been a constant feature of life on Earth, it is now proceeding at an unprecedented rate because of human activities. The most vulnerable species are endemics.

4   Habitat – an organism-centred view of the environment

4.1   Introduction

As we have already seen from the variation in vegetation which occurs across the globe, plants and animals have geographical distributions which confine them to certain types of environment or habitat. Any habitat has certain distinctive features: a particular climate, a particular kind of soil, a particular kind of water (either fresh or marine, running or still) and so on. Most species are confined to a single kind of habitat, though some species are able to live in a range of habitats, and others change habitats during their lives. For example, the Atlantic salmon spends much of its life at sea (in a marine habitat) but breeds in rivers (a freshwater habitat). Both habitats are essential for salmon to complete their life cycle.

Activity 2

While reading the three accounts of habitats which follow, you should have these questions in mind:

- What are the physical characteristics of the habitat?
- What are the most common species of plants and animals found there?
- What are the relationships between the plants and animals?
- Is the character of the habitat dependent upon human intervention, or upon its protection from human activities?

4.2   Some examples of habitats in Britain

The landscape of Britain is noted for its variety, but here we only have space to describe a few examples of the many habitats to be found, and their different dependence on, or vulnerability to human activities.

An oakwood

Seven and a half thousand years ago, woodland covered most of the British Isles. Analysis of pollen deposited in the sediments of ponds and bogs in different parts of Britain reveals the composition of the wildwood before the impact of human activities (Figure 6.9).

Today's varied landscape is the product of more than two thousand years of human influence. In southern England oaks *Quercus petraea* and *Quercus robur* now dominate the remnants of the ancient woods where the small-leaved lime *Tilia cordata* once reigned. Although only fragments remain, small ancient woodlands are not uncommon in some places such as East Anglia. Such woods invariably have a long history of forestry for timber and of management for firewood and other woodland products, but

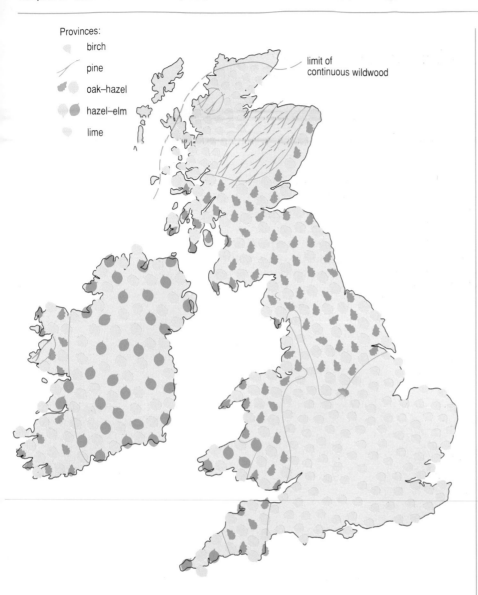

Provinces:
   birch
   pine
   oak–hazel
   hazel–elm
   lime

limit of
continuous wildwood

▲   *Figure 6.9   Wildwood provinces about 4500 BC. The diagram is simplified to show
the distribution of only the main types of woodland.*

because of their continuity they have remained a rich habitat for wildlife.
Figure 6.10 illustrates some of the species found in an oakwood.

Monks Wood in Cambridgeshire is an ancient woodland that is
reasonably typical of the remnants of wildwood in southern Britain. It is
small, only 157 ha in extent, but has been visited by naturalists for over a
hundred and fifty years. It now has the status of a National Nature Reserve,
and lies next to an ecological research station belonging to the Institute of
Terrestrial Ecology.

Pedunculate oak *Quercus robur* and ash *Fraxinus excelsior* are the most
common trees, and young trees of the latter species spring up wherever the
canopy is opened and admits light to the woodland floor. The largest oaks
have a single trunk that would have provided timber when the wood was

managed commercially. Many of the older ash trees have several stems because they have been previously cut near the base. This practice, known as coppicing, supplies a renewable source of stems for firewood or other purposes, and actually prolongs the life of the tree, which regrows after every cut. Sixteen other species of tree are also found in the wood, including silver birch, downy birch, crab apple, wild pear, wild service tree, aspen and elm. The last three of these species all produce root suckers, and they tend to occur in clumps. A stand of elm had to be felled in 1972 after an attack of Dutch elm disease, but elm stumps retain the ability to regenerate from suckers.

There are also over a dozen species of shrubs in the wood, hazel *Corylus avellana* being the most common one. In Monks Wood, as in most woods where it occurs, hazel was traditionally coppiced and its regrowth after cutting is vigorous. Few hazel seedlings are found in Monks Wood, perhaps because squirrels take most of the nuts before they ripen.

The plant species growing on the floor of the wood are dependent upon what light reaches them through the tree canopy. The commonest species is dog's mercury *Mercurialis perennis* which is able to survive in shade, and in drier places grows so abundantly that only mosses can share its habitat.

Coppicing admits light to the woodland floor and this encourages the growth of bluebell, primrose, wood anemone and wood sorrel. In coppice woods certain short-lived species such as marsh thistle, foxglove and

▲   *Figure 6.10   An oakwood, showing (not to scale) plants and animals mentioned in the text. Key: a, great spotted woodpecker; b, lichen; c, woodruff; d, nuthatch; e, bluebell; f, primrose; g, grey squirrel; h, purple hairstreak (male); i, oak leaf and acorn; j, wood anemone; k, wood sorrel; l, weasel; m, mistletoe; n, tree creeper; o, cockchafer; p, purple hairstreak (female); q, bracket fungus; r, oak wasp galls.*

mullein appear in great numbers where the canopy has recently been cut back, germinating from seed that has been lying dormant in the soil.

As well as the dormant seeds of short-lived plants, the soil harbours huge numbers of invertebrate animals which live upon dead leaves and other organic matter. 'Invertebrates' is a catch-all term for a miscellaneous group of animals that lack a backbone: indeed they lack any bones at all! Examples include insects, spiders, earthworms and woodlice. Insects called springtails are some of the commonest animals in the soil. A handful of leaf litter typically contains 70 of these animals belonging to seven different species. Larger invertebrates involved in decomposition include earthworms (12 species recorded), woodlice (7 species) and millipedes (13 species) and mites. Centipedes (9 species) and some of the 122 species of spider recorded in Monks Wood are among the invertebrates which prey upon other animals found in the soil.

Fungi also play an essential part in the decomposition of dead organic matter. About three hundred species have been recorded in Monks Wood. The conspicuous fly agaric with its red cap and white scaly spots is particularly associated with birch trees. This toadstool is actually only the fruiting body of the fungus, which has an extensive network of very fine, rootlike filaments called hyphae that ramify through the soil, gathering nutrients.

Q   You may be familiar with bracket fungi such as the birch polypore, whose plate-like fruiting bodies are often seen jutting out of decaying branches or dead logs. Where do you think these fungi obtain their food from?

A   Their hyphae ramify through decaying wood, and they obtain their food from this source.

The roots of birch trees may become wrapped in a net of hyphae of the fly agaric, and some of the mineral nutrients gathered from organic matter in the soil by the fungal hyphae are passed on to the trees. This kind of association between a fungus and a plant is called a mycorrhiza. Although plants pay a price for the minerals they obtain from the fungus, which takes sugars from their roots in return, most plants possess mycorrhizal associations and many cannot grow properly without them.

Orchids are entirely dependent upon the fungal component of a mycorrhiza during their early development, which may last for many years. Ten orchid species occur in Monks Wood, the more abundant ones being common twayblade *Listera ovata*, the common spotted orchid *Dactylorchis fuchsii*, and the early purple orchid *Orchis mascula*. The bird's nest orchid *Neottia nidus-avis* which also occurs in the wood is wholly parasitic upon its mycorrhizal fungus throughout its life.

The British orchids, though they lack the showy display of tropical species, are none the less interesting for their relationship with the pollinating insects attracted to their flowers. The green flowers of the common twayblade, which are borne on a spike projecting from the two leaves that give the plant its name, are particularly attractive to ichneumon wasps which feed upon its nectar. Ichneumon wasps parasitise the larvae of other insects by laying their eggs inside them. The ichneumon larva which hatches from the egg feeds upon the tissues of its host, which is eventually killed before it can complete its development into an adult. Parasites of this kind (called parasitoids) are important in nature because they control the numbers of insects, such as moth caterpillars, which feed upon plants. ·

Over 400 species of moths have been recorded in Monks Wood, so the possibility that plants there might be overcome by millions of little rampaging herbivores is a real one. Indeed, it does occasionally happen. In some years, on a stroll through an oakwood in late spring after the trees have leafed, you may hear a gentle but incessant pattering that sounds like light rain falling upon leaves. In fact it is raining, though the rain is not water but the faeces from millions of moth caterpillars feeding on oak leaves. The winter moth is the species most often responsible for this, but other species are prone to these outbreaks too, and in a bad year an entire oakwood may be leafless by summer.

In every sense insects are the most numerous animals in Monks Wood. Not only are there uncountable numbers of individuals, but there are also impressive numbers of species in the wood. Quite apart from the groups already mentioned, there are over 200 species of true bug, 500 species of flies, and over 1000 species of beetles. This preponderance of beetles occurs in all habitats where insects have been recorded, but just why this should be so is still one of nature's secrets.

When people talk of woodland animals, they generally think not of insects, but of mammals: foxes, badgers, squirrels, wood mice and so on. All these species and twenty others have been recorded in Monks Wood. Most woodland mammals are nocturnal, which makes their numbers difficult to guess at. A survey estimated the number of individuals of some of these species in Monks Wood in 1971 (Table 6.2).

Birds are of course the most conspicuous woodland animals. Forty-eight species breed regularly in Monks Wood, but another 67 species have been recorded flying in or over the wood. The breeding species, which are typical of deciduous woodland in lowland England, include woodcock, marsh and willow tits, nightingale, blackcap, garden warbler and great-spotted woodpecker. The numbers of different species have changed over the years as changes in the woodland have affected the suitability of the habitat for different species. In the 1920s the wood was clear felled. This

Table 6.2   The diet and estimated numbers of individuals in 1971 of some species of mammal living in Monks Wood

Species	Numbers	Diet
fox	5	rabbits, small birds, insects, small mammals
stoat	4	Small mammals, small birds, insects, bird eggs
hare	10	plants
weasel	100	small mammals, small birds, insects, bird eggs
hedgehog	150	earthworms and large invertebrates
grey squirrel	250	tree buds, shoots, fruit, nuts, bird eggs and young
rabbit	300	plants
mole	500	earthworms
common shrew	3000	small invertebrates: e.g. woodlice, springtails
pygmy shrew	500	small invertebrates: e.g. woodlice, springtails
wood mouse	5000	plants, seeds, small invertebrates
bank vole	10 000	mainly plants, some small invertebrates

Source: Steele, R. C. & Welch, R. C. (eds) (1973) *Monks Wood: a nature reserve record*, The Nature Conservancy/NERC, Monks Wood, Huntingdon, p. 289.

produced open conditions and scrub which favoured grasshopper warbler, whitethroat, linnet and yellowhammer, but as new trees grew up the habitat became less suitable for these species and their numbers decreased. Other birds which nest in holes, or which feed on invertebrates that are found in rotting wood, require a habitat containing old trees. Such species as lesser-spotted woodpecker, nuthatch and treecreeper are still uncommon from the lack of old trees in the wood. As the trees mature these birds may be expected to occur in greater numbers.

Heathland

Heathlands are defined by their characteristic vegetation, dominated by dwarf shrubs (Figure 6.11). The most ubiquitous of these plants is heather, also known as ling *Calluna vulgaris*. Its purple flowers are responsible for the lavender patchwork which blooms over heathlands in summer. The other characteristic, habitat-forming plants of heathland are the bracken fern *Pteridium aquilinum* and two species that bear a family resemblance to ling, but can easily be distinguished from it on close inspection: the cross-leaved heath *Erica tetralix* and bell heather *Erica cinerea*. Gorse bushes and birch trees are also frequent. Upland heathlands, or heather moors, are sometimes regarded as a separate type of habitat, but upland and lowland heaths will both be discussed here.

Heathlands have a reputation as wild and waste places – Egdon Heath was the fictional name Thomas Hardy gave to 'the sombre scene of the story' in his novel *The Return of the Native*, published in 1878. Hardy's native

▲   *Figure 6.11   A heathland, including some plants and animals mentioned in the text (not to scale). Key: a, silver birch; b, bracken; c, gorse; d, Dartford warbler; e, emperor moth; f, dragonfly; g, hobby; h, common bent grass; i, cross-leaved heath; j, smooth snake; k, bell heather; l, red grouse; m, heather* Calluna; *n, sand lizard.*

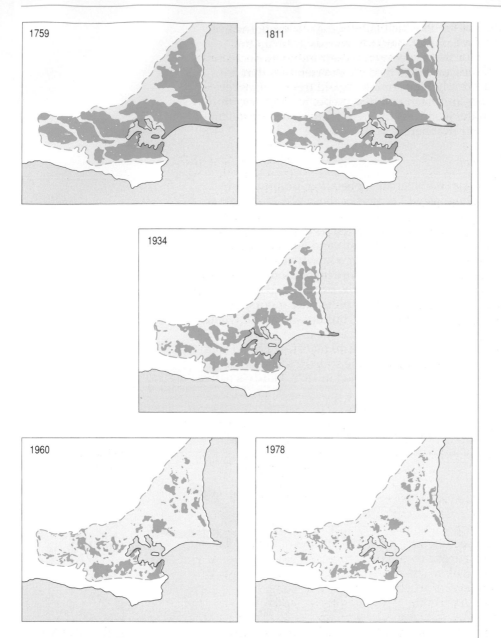

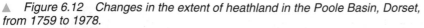

Figure 6.12   *Changes in the extent of heathland in the Poole Basin, Dorset, from 1759 to 1978.*

county of Dorset still has significant areas of heathland, though these have shrunk and fragmented greatly in the last 200 years (Figure 6.12).

The reputation of heathlands as waste places is an understandable one because this vegetation typically occurs on infertile, often sandy soil. Welsh hill farmers have a saying: 'Gold under bracken, silver under gorse, famine under heather'.

Heathlands, far from being wild, however, owe their appearance and character to human activities. Left entirely unmanaged, virtually all British heathlands below the tree-line, from Dorset to Inverness, would revert to woodland within a few decades. This kind of change from one type of vegetation to another with time is called **succession**. Given time and an accessible source of seeds of colonising species, succession tends to produce vegetation characteristic of the biome appropriate to the local climate. Where there are adult trees nearby to provide a source of seeds, heathland becomes colonised by Scots pine *Pinus sylvestris* and birch *Betula pendula* and *Betula pubescens* seedlings, which are able to grow well on heathland soils.

Human activities which may prevent this process of change on heathlands in Britain are heavy grazing, which will favour grasses at the expense of heather, shrubs and trees, and burning, which favours heather at the expense of grasses and trees. The prolonged absence of both grazing and burning favours trees. Before human intervention, it is probable that heather was a plant growing in the open areas of birch and pine woodland, where old trees had fallen. Trees and heather would perhaps have formed a patchwork, each area in turn having been relinquished by the senescence of one to the invasion of the other.

Traditional uses of heathland were for grazing cattle, sheep and ponies, and as a source of fuel. Bracken, which is toxic to sheep and cattle, was gathered for many purposes including for fuel, for animal bedding and in Scotland for thatching. Potash from burnt bracken was used to make glass and soap. Today the most important economic uses of heathland are for sheep grazing and, in the uplands, for grouse moor. The spread of bracken which has occurred in many uplands reduces the area available for sheep grazing.

The soil of heathlands is usually acid and inhospitable to the large invertebrate animals found in other soils. Consequently, earthworms and millipedes are generally absent, and mites and springtails are much less abundant than they are in the soil of Monks Wood. The absence of these animals slows down the rate at which dead plant material is broken up and decomposes in heathland soil, and a layer of plant litter builds up on its surface.

In certain heathland areas of southern Britain there are rich faunas of spiders and insects, but this plenitude appears to be due to the variety of habitats at these sites, including woodland, ponds and scrub. A diversity of insects is not characteristic of heathland elsewhere. Although the insect fauna of heather moors and heathlands in northern Britain is not great by comparison with woodland habitats, there are a number of insects that are particularly associated with heathland plants. Bracken on Skipwith Common in Yorkshire harbours species of sawflies, leaf-mining flies and moth caterpillars which eat only this plant. Bracken fronds (the leaf of a fern is called a frond) have nectaries at their base which exude a sugary secretion; this attracts insects to the plant, where the ants, ladybird beetles and parasitoid wasps prey upon the herbivores feeding there.

Heather also has its own particular insects associated with it. The flowers of *Calluna* and *Erica* harbour thrips: insects so tiny that they may live six to a flower (Figure 6.13). Most of the insect's life is spent within the flower, feeding upon pollen, but when mature the thrips squeezes its way out, carrying away pollen on its body as it leaves. After mating, the female enters a flower on another plant to lay her eggs, depositing pollen on the flower's stigma on her way in, and so effecting cross-pollination between bushes.

About thirty species of larger moths eat *Calluna*, seven or eight of them eating nothing else. Two of these specialists, whose common names betray the romantic imagination of Victorian moth collectors, are the true-lover's

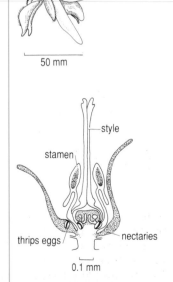

50 mm

style

stamen

thrips eggs   nectaries

0.1 mm

▲   *Figure 6.13*
A cross-section of a Calluna
flower showing a heather
thrips and its eggs.

knot and the beautiful yellow underwing. The caterpillars of the latter species closely resemble a *Calluna* shoot. One of the insects which eats other heathland plants as well as *Calluna* is the caterpillar of the emperor moth *Saturnia pavonia*, which is the only native British representative of the silk-moth family. Females fly only by night, but males can be seen flying over the heather by day. In August the larva spins the silken cocoon of the pupal stage among heather twigs, and emerges as an adult in the next spring.

An important consumer of *Calluna* is the heather beetle *Lochmaea suturalis* whose larvae and adults all feed on the plant (Figure 6.14). In the north of England, and on the continent of Europe, outbreaks of the heather beetle can cause defoliation of *Calluna*, severe enough to kill the bushes. In recent years defoliation by heather beetle appears to have become more frequent on heathlands in the Netherlands where nitrate deposited from atmospheric pollution has a dual effect upon the vegetation. The nitrate fertilises the soil, to the benefit of grasses, whose vigorous growth causes increased competition with heather. *Calluna* also takes up additional nitrate from the soil, but this raises the food quality of its leaves for the voracious heather beetle, whose populations increase accordingly. It is believed that atmospheric pollution by nitrates has led to a decline in heather and the conversion of former heathlands to grasslands through a combination of increased competition from grasses and the weakening of *Calluna* by frequent defoliation.

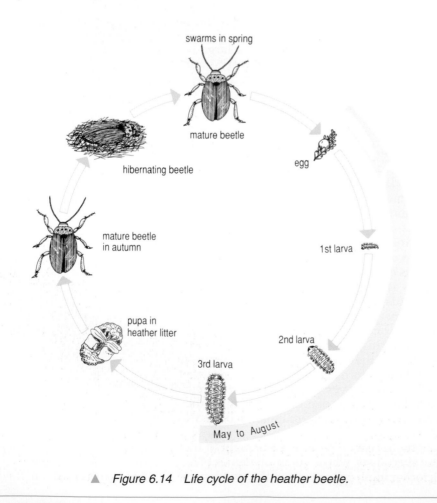

▲   *Figure 6.14   Life cycle of the heather beetle.*

Even when left unmolested by beetles, heather bushes have a natural cycle of growth, maturity and decay (Figure 6.15). As they grow older, *Calluna* bushes produce less and less new foliage. Old bushes lose their compact outline, become 'leggy' and their stems collapse sideways, leaving a gap in the centre of the bush that can be colonised by other plants. If allowed to proceed to a natural conclusion, this process would lead to the disappearance of *Calluna* and its eventual replacement by scrub and then woodland.

Like the Phoenix of legend, *Calluna* can be rejuvenated by fire. A rapid burn, which passes over the ground too quickly to penetrate the soil, stimulates regrowth from old *Calluna* roots and triggers the germination of *Calluna* seeds from the soil, so that within a few months of an autumn or spring burn, new *Calluna* shoots and seedlings appear. Within a few years of a burn, the heathland is verdant with a uniform, lush growth of young heather. Deliberate, controlled burning of heathland in patches is used to manage grouse moor and other heathlands. Apart from rejuvenating the heather, it also kills the birch and other tree seedlings which constantly threaten to invade this habitat and turn it to woodland.

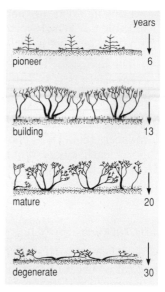

▲ *Figure 6.15*
The life cycle of a Calluna *bush.*

Q   Why should heather moor be burnt in patches, and not in whole large areas at once?

A   Although the heather is able to survive fire, invertebrate animals are not, though they can quickly reinvade burnt areas if some neighbouring areas are left unburnt for a time.

On lowland heaths, where there is now little grazing by livestock, rabbits can make a significant impact upon the vegetation by grazing *Calluna*, gorse and bell heather. They avoid eating cross-leaved heath and bracken. Apart from sheep, the principal herbivore on Scottish upland heathlands is red grouse *Lagopus lagopus* subspecies *scoticus*. Adult red grouse feed on new, green heather shoots, and their young chicks live on insects collected from the bushes, so grouse are wholly dependent upon *Calluna*. New green shoots are most abundant on young heather, but older, larger bushes are required as shelter for the birds. This is obviously another reason why a heathland managed for grouse should have areas of old as well as young heather. A heathland managed only for sheep on the other hand provides most grazing for the animals if there are no unproductive old bushes.

Apart from the red grouse, which is confined to upland heath and moor, there are few birds that are found exclusively on heathland. Most of the species found there are woodland birds. A notable exception is the Dartford warbler *Sylvia undata* which reaches the northern extremity of its geographical range in the south of England. The Dartford warbler depends upon a year-round diet of insects, and forages for beetles, spiders, caterpillars and bugs on gorse bushes, which are therefore an essential component of its habitat on all heathlands where it occurs.

A bird of prey particularly associated with lowland heath is the hobby *Falco subbuteo*. It is a summer visitor to Britain that prefers to nest in trees standing in open land: hence its occurrence on heathland. The hobby preys upon large insects and dragonflies, which are found coursing to and fro over heathland pools, and upon swallows and martins.

The mammals found on heathland are not peculiar to this habitat, and are characteristically species of woodland. The scarcity of large soil invertebrates limits the food available for some woodland mammals which are consequently rare or absent from heathland.

Q   Look again at the diets of the woodland mammals recorded in
Table 6.2. Which species that feed upon large soil invertebrates would
you expect *not* to be found in heathland habitats?

A   The hedgehog and mole feed upon earthworms and other soil
invertebrates, so you should not expect to find them in heathland
habitats.

The famous Scottish painting 'The Monarch of the Glen' by Sir E. H.
Landseer (1802–1873) shows a red deer stag standing at bay upon a heather
moor, clearly suggesting that this animal is a proud native of upland heath.
However, this species, *Cervus elaphas*, is chiefly a denizen of woodland,
though it grazes on open moorland in the summer.

The rocky seashore

A rocky seashore is probably the nearest one can come in Britain to a
habitat where human activities are *not* the dominant influence upon its
appearance and its inhabitants. It is also a remarkable habitat for another
reason: the animals and plants in it are arranged in a much more regular
fashion than can be found in other habitats.
    Figure 6.16 shows the distribution of algae and invertebrate animals
typical of rocky seashores on the west coast of the British Isles.

Q   *Chthamalus stellatus* and *Balanus balanoides* are two different species of
barnacle. How are they distributed relative to each other in this habitat?

A   *Chthamalus stellatus* occurs in a band at the top of the shore and *Balanus
balanoides* occurs in a band below it.

The kind of distribution shown by the two barnacles is typical of the other
organisms in this habitat, which all occur within broad bands across the
shore. This banding pattern is called zonation. The obvious question which
will probably occur to you is why do the plants and animals occur in bands
like this? The brief answer is that there is a gradient in the physical
environment down the shore and species respond differently to this.
Physical disturbance from waves and the period of exposure to the air vary
greatly between high and low water marks.
    The gradient in these physical factors can explain some of the zonation
in the distribution of species. For example, *Chthamalus stellatus* is more
resistant to desiccation than *Balanus balanoides*, which is why it can occur
higher up the shore. But biological factors are responsible for the pattern
too. The upper distribution limit of *Balanus balanoides* might be determined
by the hazard of desiccation high up the shore, but what determines the
lower limit of the *Chthamalus stellatus* band? Although as adults barnacles
are stuck to the rock, as tiny free-floating larvae they are washed to and fro
across the entire shore by the tides. So it is reasonable to ask why adult
barnacles don't occur lower down. A reasonable hypothesis is that the two
species of barnacle compete for space, and that perhaps the presence of
Balanus balanoides is responsible for the lower distribution limit of
Chthamalus stellatus. An ecologist tested this hypothesis on the shores of
Scotland by simply removing *Balanus balanoides* from parts of the shore and
returning to find out what happened some time later. He found that
Chthamalus stellatus had invaded the top of the *Balanus balanoides* zone, but
only where the latter was removed.

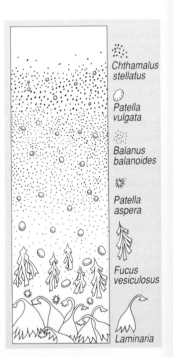

▲   *Figure 6.16*
The distribution of
Chthamalus *and* Balanus
*barnacles on a semi-
exposed shore.* Fucus
vesiculosus *and* Laminaria
are seaweeds, Patella
vulgata *and* P. aspera *are
limpets.*

Many biological factors vary in parallel with the physical gradient, thus reinforcing its effect upon the distribution of animals and plants found there. For example, shore birds search for food behind the retreating tide. The higher reaches of the shore will be more dangerous for their food animals because these areas are exposed for a longer period each day. Conversely, animals in these same areas will be less vulnerable to marine predators. For example, the dogwhelk *Nucella lapillus*, which feeds upon submerged barnacles, can spend more time feeding on the lower shore than on the upper. Other predators such as crabs and fish may not be visible to us when we visit the seashore at low tide, but will also affect the distribution of their prey.

Grazing by limpets and marine snails is a significant factor in determining the distribution of algae upon the seashore. When the tide is out, limpets clamp themselves immovably to a rock and the casual observer would have little idea of the part these animals play in the seashore community. You will sometimes see a pattern of radiating lines in fanlike clusters around a stationary limpet, which is one clue to what happens when the tide comes in and the limpets become active. You might also notice that although the rock around a limpet is bare of algae, some of these animals have a little tuft of algae perched upon their shells. This is another clue.

What is the significance of these observations? The fanlike clusters of lines are the toothmarks left by the grazing activity of limpets when covered by the tide. When submerged, limpets creep around the rocks scouring off the scum of algae which grows upon them. This prevents large algae from establishing. Limpets may also crawl over each other's backs grazing algae from them, but when limpet densities are low a limpet's shell may be the only place where a limpet's tongue never reaches: hence the little tuft of algae.

Limpets are not very big, and it is easy to underestimate their impact, even though they occur in large numbers. Curiously enough, human interference with the rocky shore can provide the occasion that demonstrates the power of limpet and snail grazing. In 1967 a major pollution incident affected the rocky shores of Cornwall when an oil tanker, the *Torrey Canyon*, was shipwrecked and spilled its cargo into the sea. The oil which washed up onto beaches was sprayed with detergent to try to disperse it, and this killed all the molluscs on the shore. When the oil and detergent had gone, the shore became covered in a luxuriant growth of the alga *Fucus*, the like of which had not been seen before. The *Fucus* inhibited the settlement of limpet and barnacle larvae on the shore in the following year, and it was five years before limpets became established again.

Activity 3

You should now try the following questions to help you understand the links between organisms in each of the three habitats which have been described.

Q1 Name one example from each habitat in which grazing by an invertebrate animal can have a significant impact upon the main plant species there.

Q2 Name one example from each habitat in which human activities have *indirectly* altered the flora or fauna.

Q3 Name one example from the woodland habitat and one example from the heathland habitat in which the presence or absence of a species can be ascribed to the presence or absence of another organism that consumes dead organic matter.

Q4 Name one example from the heathland habitat and one example from the rocky shore habitat in which competition between species affects their distribution.

Q5 One kind of relationship between plants and animals is the one in which plants provide food for herbivores. Name two other ways in which plants and animals may be dependent upon each other, as described in any of the three habitats.

4.3   Summary

The term 'habitat' is used to describe the *kind* of environment in which an organism lives. The physical conditions of soil and climate as well as the other animals and plants found there may all be essential to creating a habitat suitable for a particular species. Habitats are often named after the physically dominant organisms present there – for example 'oakwood' or 'heathland' – but there are generally many other species present too, some of which may seem insignificant but which have vitally important relationships with the dominant organisms. The mycorrhizal fungi in an oakwood or the heather beetle in heathland are examples. Relationships between the different organisms within a habitat may be complex and diverse and they require detailed ecological study to untangle.

Human activities influence virtually all habitats in Britain, and many habitats (such as heathland or coppice woodland) owe their existence almost entirely to human intervention or management. The rocky seashore is perhaps the habitat least shaped by management, though when human interference does occur it can have surprising effects because of its impact upon the unseen relationships between animals and plants.

5   Conclusion and summary

The global pattern of vegetation on the Earth's land surface is divided into biomes, which reflect the relationship between climate and plant life. Plants provide the living framework for other organisms. We share this planet with an uncounted number of other species, which may number as many as 30 million in all. Arguments for the conservation of these species include their practical value to humans and the moral view that other species have a right to exist. Regardless of which attitude is adopted, the conservation of species for their own or for our sake needs to be based upon a knowledge of what exists, in what quantity, and where it is to be found.

Putting the uncountable numbers of micro-organisms to one side, we find that among described species the most numerous group is the insects, and among insects the beetles, which comprise fully one quarter of *all* named species. Estimates of the number of species on Earth as a whole suggest that the bulk of invertebrate animals have not yet been described. For many groups, the largest numbers of species are found in the tropics where the habitats that harbour them are under threat.

Habitat destruction and hunting have caused the extinction of many species of vertebrates, and estimates of modern extinction rates suggest that the process now occurs at a rate unprecedented in the Earth's long history. To protect species from extinction, the habitats in which they are found must be protected, because all species depend to a greater or lesser degree upon relationships with other animals and plants.

The ecological interrelationships we have encountered in a British oakwood, in heathland, and on the seashore are but a small taster for the richness of ecological systems. These examples show that human intervention can have surprising and unpredictable effects upon habitats and their fauna and flora, but that some habitats, such as heathland, may depend upon human activities for their very existence. In order to protect habitats and the species they contain more effectively, their ecology must be better understood.

In the next chapter we will look at ecological systems in more detail, to see how they function. On this functioning, and upon our understanding and wise management of it, depend the future of habitats for countless species.

Answers to Activities

Activity 1

Many of the examples given in Table 6.1 were hunted to extinction. While this still threatens many big animals, for example the African elephant, habitat destruction is probably the biggest threat to wildlife as a whole today. Habitat destruction can be direct, by the pollution of lakes, or by the physical removal of forest for example, but habitats can also be altered radically by the introduction of alien species. The Nile perch in Lake Victoria is an example of the latter.

Rare species or the inhabitants of threatened habitats are obviously at risk. Species endemic to small areas such as oceanic islands are the most vulnerable of all.

Activity 2

In the woodland, and where trees are present on heathland, the trees themselves strongly influence one of the most important physical characteristics of these habitats, namely light conditions on the ground. On the rocky seashore, inundation by the tides and desiccation are dominant physical characteristics of the habitat.

Monks Wood has been managed for timber and fuel for centuries, and its present character has been shaped by this. Human intervention alters the composition of the flora and fauna in the wood. If left to itself the habitat might become less suitable for some species and more suitable for others, but it would not disappear altogether. Heathlands are much more dependent upon human activities. The balance between fire and grazing can alter the habitat entirely, causing a change from heathland to grassland, or allowing succession to woodland.

Activity 3

Q1   *Woodland*: defoliation of oak trees by winter moth. *Heathland*: defoliation of *Calluna* by heather beetle. *Rocky seashore*: prevention of algal establishment by limpets.

Q2   *Woodland*: The change in bird species with maturity of the woodland in Monks Wood was an indirect result of the clear felling earlier in the century. *Heathland*: Atmospheric pollution in the Netherlands has led indirectly to a decline in heather and heathland habitat through its effect upon grass growth and heather beetle nutrition. *Rocky seashore*: The spraying of beaches with detergent after the *Torrey Canyon* disaster led indirectly to the colonisation of rocks by algae after limpets were killed.

Q3   *Woodland*: Orchids, particularly the bird's-nest orchid, depend upon mycorrhizal fungi. Hedgehogs and moles eat earthworms, which are important in litter decomposition. *Heathland*: Hedgehogs and moles are virtually absent because of the absence of earthworms and other large soil invertebrates from heathland soils.

Q4   *Heathland*: Competition between grasses and *Calluna* can affect the distribution of the latter species. *Rocky seashore*: Competition between *Chthamalus* and *Balanus* barnacles determines the lower limit of *Chthamalus* on the shore.

Q5   Insect pollination of flowers, e.g. ichneumon wasps pollinating twayblade in the wood; thrips pollinating heather in the heathland. Plants providing shelter for animals, e.g. hole-nesting birds in the wood; grouse sheltering in old heather and the hobby nesting in isolated trees in heathland.

1   Introduction

In Chapter 6 we began our inventory of the biosphere by looking at the diversity of life on the planet as a whole, and how this is apportioned around the globe. For life on land, the biosphere can be divided up into the major biomes, which represent distinctive types of vegetation associated with particular types of climate (Plates 4–10). This could be thought of as an external view of the biosphere – how things would appear from a satellite. We then changed our perspective, and took an internal view of the biosphere, looking at a small selection of habitats.

Biome and habitat are descriptive terms appropriate to the task of Chapter 6, which was mainly to sketch the inhabitants of the biosphere. The organism-centred view of the environment represented by the concept of habitat revealed that many species are linked to one another in ecological relationships. Many of these links are highly specific, as for example the relationship between the heather beetle and *Calluna*. We saw in this particular case that a disturbance of the relationship can have far-reaching consequences.

This chapter will now concentrate on ecological relationships to see how the organisms found in habitats *function* together. To do this we will look at two levels of ecological organisation: the **ecosystem** and the **population**. The term 'ecosystem', introduced in Chapter 6, is used for an ecological system composed of interlinked species and their physical resources. Note that this term introduces and emphasises the idea of *links*. The term 'habitat' is confined to describing where and with what other species an organism lives. The idea of an ecosystem is concerned with how the relationships between organisms work. We will first look at some general properties of all ecosystems and then look separately and in more detail at terrestrial (land-based) and marine ecosystems.

The second level of ecological organisation examined in this chapter is that of the population. This is defined as a collection of organisms belonging to the same species, found within a limited geographical area. Looking at populations and how they work gives us an even more detailed insight into ecological relationships, as you will discover in the later sections of this chapter.

As you read this chapter, look out for answers to the following key questions.
- How are different kinds of organism linked by the flow of energy and cycles of minerals in ecosystems?
- How are population growth rates analysed?
- Why does genetic diversity matter?

2   Ecosystems

2.1   Introduction

In this section we will begin by constructing a model ecosystem which has the essential features common to all real ecosystems.

Q   Remember the basic biological fact that all living things require a source of energy, and that this ultimately comes from the sun. What group of organisms would you expect to find in every ecosystem, whether in the sea or on land, whether in a lake or on a mountain top?

A   Green plants. Only green plants and a few bacteria can turn sunlight into chemical energy by photosynthesis.

Plants are the foundation stone of our model ecosystem. The quantity of carbon compounds that they fix into their tissues each year is called primary production, and the plants themselves are referred to as **primary producers**. Phytoplankton are the main primary producers in marine ecosystems, trees fill this role in forests, and wheat (and maybe weeds!) are the primary producers in a cornfield.

Q   Because of their ability to 'feed themselves', plants are autotrophes. What other class of organism would you expect to find in the model ecosystem?

A   Heterotrophes.

Heterotrophes may be divided into animals which eat plants (called **herbivores**), animals which eat animals (**carnivores**), and fungi, bacteria and animals which live on dead plants and animals (**decomposers**).

We now have all the components of the model ecosystem ready for assembly: primary producers, herbivores, carnivores and decomposers. Each of these components of the ecosystem is represented by a box (called an **ecosystem compartment**) in the diagram shown in Figure 7.1. An extra box is labelled 'dead organic matter' to represent the dead remains of living organisms which have not (yet) been consumed by decomposers. This compartment represents, for example, the dead heartwood standing in the trunks of living trees and the peat which accumulates in bogs.

Activity 1

Using what you know about the energy sources used by each component of the ecosystem, draw arrows between the boxes in Figure 7.1 to trace the flow of energy through the model ecosystem.

Then turn to Figure 6.10 in Chapter 6 which shows some of the organisms you might expect to find in a British oakwood. Select examples from Figure 6.10 of a primary producer, a herbivore, a carnivore and a decomposer and write the names of these in the appropriate boxes of Figure 7.1.

When you have done this, check your answer with Figure 7.21 near the end of this chapter.

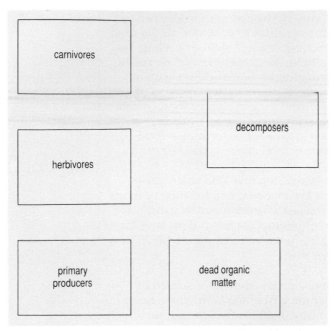

▲   *Figure 7.1   The compartments of a 'model' ecosystem.*

The organisms in each compartment of the ecosystem use energy in their respiration. This energy is lost as heat from all organisms, including plants. In plants, only a proportion of the sun's energy they capture is incorporated into their tissues and contributes to primary production.
In heterotrophes, only a proportion of their food is used in growth. So, as energy flows from plants to herbivores, from herbivores to carnivores, and from each of these to decomposers, there is a loss at every stage and the amount of energy available to the recipients becomes less and less.
   Figure 7.2 is a simplification of the feeding relationships between organisms, and is called a **food chain**. As well as the inevitable losses *within*

▲   *Figure 7.2*
A simple, hypothetical food chain showing how energy is lost by respiration at every step.

each compartment due to respiration, some energy also goes astray *between* links in a food chain because animals are never able to consume all the food potentially available to them. This is not a loss to the system as a whole (as respiratory heat can be) because, for example, the grass not eaten by rabbits will instead pass to decomposers or to other herbivores such as sheep or caterpillars. In fact, because most organisms in an ecosystem depend upon more than one other species for their food, and may in turn be food (when alive or after death) for more than one species, it is more accurate to describe the feeding relationship between organisms as forming a **food web**. It should be remembered that the term 'food chain' is used only as a convenient simplification.

The amount of energy passing along each step of a food chain varies greatly depending upon what organisms are involved and the type of food they eat. In marine ecosystems herbivorous zooplankton may consume as much as 40% of the primary production of phytoplankton, but on land the insects which eat plants rarely consume more than an average of 10% of the primary production in a terrestrial ecosystem.

Q   Assuming that only 10% of the primary production is consumed by herbivores, and that carnivores consume 10% of the herbivores, how much of the original energy would be available to feed an animal which preyed upon the carnivores, taking 10% of them?

A   10% of 10% of 10%, which is one tenth of one per cent.

This simple calculation has some interesting and far-reaching consequences. Certain animals, like the polar bear, eat other carnivores and must survive on a food supply that can never exceed a tiny fraction of the primary production of the ecosystem in which they live. Such animals, called **top carnivores** because they are at the top of the food chain, must be large in order to hunt and overcome their prey, so they have a correspondingly large demand for energy. A large appetite for relatively scarce food can only be satisfied by hunting over large areas of habitat. This not only makes such animals rare but also makes them very vulnerable to the shrinkage of habitat which is threatening all wild areas.

2.2   Decomposition

Notice that the arrows representing energy flow between the compartments of the model ecosystem in Figure 7.1 all end up with the decomposers. In terrestrial ecosystems exactly how much energy remains locked up in dead organic matter and how much of this organic matter is broken down by decomposers, to be finally dissipated as heat, depends upon conditions in the soil where decomposing organisms do their work. In ecosystems where the soil is cold during much of the year (as in the arctic), dry (as in deserts), highly acid (such as in heathlands) or lacking in oxygen due to waterlogging (such as in peat bogs), decomposition by bacteria and other organisms and the flow of material between the organic matter compartment and the decomposer compartment is weak. This causes undecomposed dead organic matter to accumulate. The most important consequence of this is that the minerals contained in the organic matter remain locked up and unavailable to plants.

You should now try Activity 2, which is about a blanket bog ecosystem (illustrated in Plate 11) at Moor House National Nature Reserve in the Pennines, where soil conditions do not favour rapid decomposition.

Activity 2

At Moor House National Nature Reserve the peat reaches a depth of 2–3 m. The climate at this site is cold, wet and windy, with ground frost occurring in almost every month of the year. Moor House is at an altitude of between 520 m and 642 m above sea-level, and is 100 m above the tree-line in this part of Britain. The main plant species are heather *Calluna vulgaris*, cotton grass *Eriophorum* spp. (i.e. more than one species of this genus), and bog mosses *Sphagnum* spp. Sheep find poor grazing here, and the other herbivores are grouse and plant-sucking bugs. The only carnivores are spiders and a few birds. Lapwing, curlew, snipe, redshank, golden plover, dunlin and red grouse all breed on the reserve. The soil fauna which feed on plant material include the larvae of crane flies (tipulids), a group of earthworms tolerant of cold conditions known as enchytraeids, and various mites, roundworms (nematodes) and beetles.

Figure 7.3 is a simplified diagram of energy flow for the blanket bog ecosystem at Moor House.

Q1   What proportion of annual net primary production is consumed by herbivores at Moor House?

Q2   In terms of the amount of carbon consumed, how important is the soil fauna compared with the herbivores?

Q3   Although the soil fauna consumes 40% of the primary production, notice that most of this (120 grams of the 131 g of carbon consumed) is returned to the pool of dead organic matter in their faeces. What *finally* happens to most of the primary production?

Despite this considerable activity by decomposers in the very wet, cold conditions, decomposition is slow and about 10% of the carbon fixed by plants each year accumulates as peat.

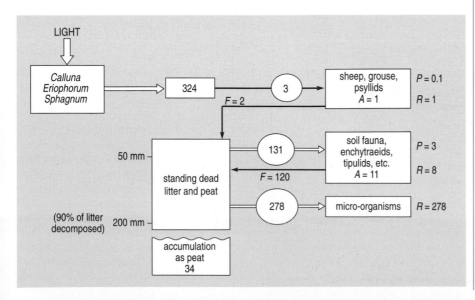

▲   *Figure 7.3
Energy flow for the blanket bog ecosystem at Moor House National Nature Reserve on the Yorkshire Moors, northern Pennines. Values (in grams of carbon per m² per year) in boxes are for net production, those in circles are for consumption; A, assimilation; P, production; R, respiration; F, faeces.*

2.3   Mineral elements

SIR TOBY BELCH.   Does not our lives consist of the four elements?
SIR ANDREW AGUECHEEK.   Faith, so they say – but I think it rather consists
of eating and drinking.

(William Shakespeare, *Twelfth Night*, Act II, Scene 3)

If *Twelfth Night* were written by a modern playwright, Sir Toby would be a
chemist and Sir Andrew an ecologist. We, and other animals, obtain the
mineral elements essential to life by eating and drinking.

Plants obtain their nutrients from the soil and, to a lesser though
sometimes significant extent, by absorption from rainwater falling on their
leaves. The pathway of mineral elements such as nitrogen, phosphorus,
potassium and sulphur in the model ecosystem is shown in Figure 7.4.

Q   What is the chief difference between Figure 7.1 (completed version) and
Figure 7.4?

A   There is an arrow indicating flow from the decomposers to the primary
producers in Figure 7.4 and not in Figure 7.1.

Whereas *energy flows* through the ecosystem in Figure 7.1, the extra arrow
in Figure 7.4 shows that *minerals cycle* within the ecosystem. The distinction
between the *flow* of energy and the *cycling* of minerals is an extremely
important one.

The cycling of nitrogen, phosphorus, potassium, sulphur and the more
minor elements that are essential to life (the **mineral cycle**) can be analysed
in the same way that we dealt with the carbon cycle in Chapter 5. The most

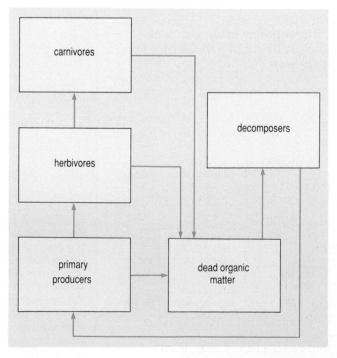

▲   *Figure 7.4*
The pathway of mineral elements in the model ecosystem.

important question for the supply of all of them is to determine their major reservoirs. The major reservoir for all the elements except nitrogen is in rocks. Minerals are slowly released to the overlying soil during the process of weathering of rocks and rock fragments.

Q   Where is the major reservoir of nitrogen in the **nitrogen cycle**?

A   In the atmosphere, which is 80% nitrogen gas (Chapter 5, Table 5.1).

Atmospheric nitrogen gas can only be used as a source of nitrogen by a few groups of prokaryotes that include blue-green bacteria, and some particular kinds of bacteria which occur mainly in association with the roots of certain plants. All these bacteria 'fix' nitrogen from the atmosphere into a chemical form that is usable by plants, and by the animals which eat them, and are known as **nitrogen-fixing** bacteria.

Although energy *flows* through ecosystems and minerals *cycle* through them, the two processes are intimately linked.

Q   We have already come across an example of the linkage between energy flow and mineral cycling, in which an obstacle to energy flow slowed down mineral cycling. What was this?

A   When decomposition is slow, organic remains accumulate in the 'dead organic matter' compartment and lock up minerals.

In the Moor House ecosystem the poor conditions for decomposition, the low nutrient content of peat, and the depth of the bedrock make the soil a poor source of mineral elements for the plants in the blanket bog. Rainfall is high (around 1850 mm a year), and this is the main source of all mineral nutrients except nitrogen, about 45% of which comes from the rain and the remainder from nitrogen fixation.

On land, decomposition occurs mostly in the soil and it is a vital process to the cycling of all minerals because it can determine how rapidly these essential elements become available again for plants and animals. The fertility of soils and the potential for sustainable agricultural production depend upon decomposition.

2.4   *Summary*

A model ecosystem, containing all the essential components of real ecosystems, can be divided into compartments representing primary producers (plants), herbivores, carnivores, decomposers and dead organic matter. The pathway of energy between plants and carnivores is called a food web. Losses of energy via respiration occur at each link in the chain, and energy flows through the ecosystem.

Mineral elements cycle in the ecosystem. The rate at which they do this depends upon the rate of decomposition which releases them from dead organic matter, making them available again to plants. Where decomposition is inhibited by wet or acid soil or by cold, minerals become locked up in undecomposed organic matter. In blanket bogs such as at Moor House dead organic matter accumulates as peat.

The main reservoir in most mineral cycles is in the rocks, but for nitrogen it is the atmosphere. Nitrogen-fixing bacteria are able to turn gaseous nitrogen from the air into a form that can be taken up by plants.

3   Soils and terrestrial ecosystems

3.1   Introduction

The soil is a world in miniature that exists at the interface between the geosphere and the biosphere. Any deep hole will reveal a soil profile that shows its history. At the bottom of the profile is the **parent material**, which may be rock, or rock debris deposited by rivers (alluvium) or glaciers (glacial drift) when they melted. Most of the better agricultural soils in Britain are found on glacial drift or alluvium. The weathering of the parent material provides the mineral matrix from which the soil is formed. At the surface of the soil profile are dead organic remains; these fuel the soil's inhabitants (Figure 7.5). Soil animals, bacteria and fungi break down organic matter and incorporate it into the mineral fraction of the soil, creating the intimate mixture of the two which is commonly referred to as topsoil.

Q   What soil organisms have we already encountered in the habitats described in Chapter 6, and at Moor House?

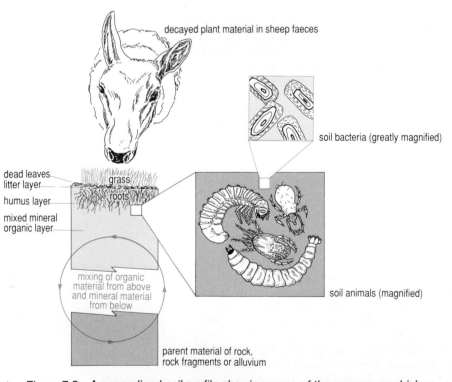

decayed plant material in sheep faeces

soil bacteria (greatly magnified)

dead leaves
litter layer
grass
humus layer
roots
mixed mineral
organic layer

mixing of organic
material from above
and mineral material
from below

soil animals (magnified)

parent material of rock,
rock fragments or alluvium

▲   *Figure 7.5   A generalised soil profile showing some of the processes which contribute to soil formation and composition. Dead organic material arrives at the surface where it forms a layer of litter. Decomposing organisms break large fragments of litter into smaller ones, and as decomposition continues the material becomes incorporated into the humus layer. If this decomposition is impeded this is the layer in which peat accumulates, otherwise soil animals such as earthworms cause organic material and mineral soil to become mixed.*

A   Mycorrhizal fungi, springtails, earthworms, centipedes, millipedes, woodlice, crane fly larvae, roundworms, enchytraeid worms, beetles and mites.

There are several stages in the incorporation of dead plant material into the soil, and at each stage the fragments become smaller and smaller, and provide the food for a different set of species in the soil fauna. The final decomposition of cellulose and lignin, which are the structural materials that contain most of the plant's carbon, can only be performed by bacteria and fungi.

In temperate ecosystems **earthworms** are important in the first stages of decomposition of dead plant material, and in the burying of organic matter. Charles Darwin made an extensive study of the activities of earthworms and observed that the species which form casts (not all species do this) can bring significant amounts of soil to the surface. In agricultural ecosystems, earthworms are most abundant and of most importance in grassland because they aerate the soil with their burrows, as well as incorporating dead organic matter such as leaves into it. In arable land earthworm activity is of less importance because the plough aerates and mixes the soil, but the smaller animals, bacteria and fungi which decompose organic matter are still important.

The ability of earthworms to change soil structure has been impressively demonstrated at Rothamsted Experimental Station in Hertfordshire, where agricultural research has been carried out for over 130 years. A grassland over 300 years old that was studied there had ten times the earthworm activity of a seven-year-old pasture, and fewer stones and a less dense surface soil than younger grasslands with fewer earthworms (Figure 7.6). Note that it is impossible to *prove* that earthworms were responsible for the progressive burial of stones, as the evidence in Figure 7.6 seems to show, but these animals bring considerable quantities of fine soil to the surface in their casts. In the 300-year-old grassland this quantity amounted to the equivalent of 6% of the soil in the top 100 mm every year.

Q   If earthworms brought 6% of the top 100 mm of the soil to the surface each year, how long would it take them to bury a stone lying on the surface to a depth of 100 mm?

PUNCH'S FANCY PORTRAITS.—No. 54

CHARLES ROBERT DARWIN, LL.D., F.R.S.

IN HIS *DESCENT OF MAN* HE BROUGHT HIS OWN SPECIES DOWN AS LOW AS POSSIBLE—I.E., TO "A HAIRY QUADRUPED FURNISHED WITH A TAIL AND POINTED EARS, AND PROBABLY *ARBOREAL* IN ITS HABITS"—WHICH IS A REASON FOR THE VERY GENERAL INTEREST IN A "FAMILY TREE." HE HAS LATELY BEEN TURNING HIS ATTENTION TO THE "POLITIC WORM."

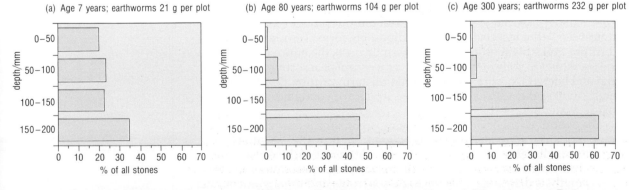

▲   *Figure 7.6   The effect of earthworms in sorting out soil and stones on grassland of different ages at Rothamsted.*

A   6% of 100 mm is 6 mm a year, so it would take 100/6 = 16.7 years. This is not very long at all, and it supports the idea that earthworms could be responsible for the difference in the distribution of stones in the soils shown in Figure 7.6.

The soil is home to harmful organisms as well as to ones that are beneficial. Among the most serious of soil pests in agriculture are the microscopic eelworms or soil nematodes which feed upon plant roots. The soil population of the eelworm which attacks potato builds up to such a level after several crops have been grown in the same ground that potatoes cannot profitably be grown in the same place year after year. Soil nematodes are not without their own enemies in the soil. The mycelium of the soil fungus *Dactylella bembicodes* has tiny traps arrayed along it, like a series of deadly nooses strung out along a washing line. When a nematode blunders through one of these traps, the noose tightens, holding it fast, while the fungus penetrates it with fine filaments (hyphae) that digest the prey (Figure 7.7).

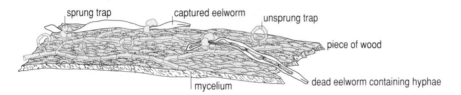

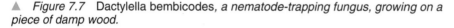

▲   *Figure 7.7*   Dactylella bembicodes, *a nematode-trapping fungus, growing on a piece of damp wood.*

Although earthworms are present in tropical soils, their importance is outweighed by a group of insects, related to cockroaches, which are confined to the tropics – the termites. Like most other animals, termites are unable to digest cellulose and lignin themselves, but they harbour in their guts tiny animals called flagellates that do the job. In the commonest group of termites, the flagellates in their guts can account for up to two-thirds of the body weight. The flagellates ingest fine particles of wood and, together with bacteria also found in the termite gut, are responsible for digesting a large proportion of the cellulose which the termite consumes.

Another group of termites feeds on plant material in a different way, by the cultivation of fungi. Inside their nests, called termitaria, the termites use their faeces to build combs; these are colonised by fungi that break down undigested plant remains. The fungi belong to species that are thought to be particularly associated with termites, and they convert food that is indigestible to the insects into a more digestible form. Termites consume old combs and fungal tissue. In these termite nests the entire decomposition process, which in soils normally involves a complex assortment of fauna, fungi and bacteria, occurs wholly within the termitarium. Dead plant material enters the nests, and the only outputs are young termites, carbon dioxide, water and heat.

The overall efficiency of decomposing organisms in recycling minerals within terrestrial ecosystems can be judged from how the essential elements are distributed between the soil and the rest of the ecosystem. Figure 7.8 shows how nitrogen, one of the most important minerals for plant growth, is distributed between the soil, roots and the organic matter (**biomass**) above ground in different ecosystems.

Activity 3

In terrestrial ecosystems most decomposition occurs in the soil. From Figure 7.8 what trend can you see in the proportion of total nitrogen present in the soil as you go from ecosystems at high latitude towards the equator?

The geographic variation in the distribution of nitrogen with latitude is explained by the role of decomposition in mineral cycling. As we have already seen, decomposition is slow in cold climates, so organic matter and nitrogen spend a longer time in the soil of ecosystems at high latitude than near the equator. This bottleneck in mineral cycles causes a concentration to build up in the soil. Although other minerals follow the pattern shown

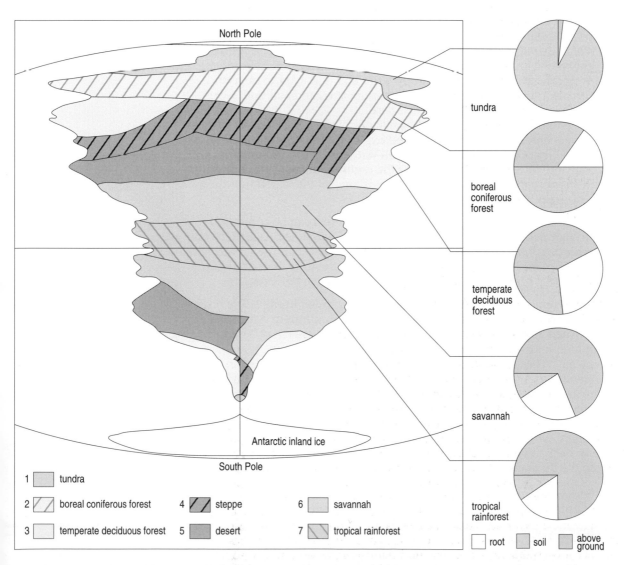

▲ *Figure 7.8   The distribution of nitrogen between the soil, roots, and the organic matter above ground in different ecosystems.*

by nitrogen in Figure 7.8, they have a reservoir in rocks rather than in the atmosphere.

The high latitudes were glaciated in geologically recent time, so soils in these areas are relatively young, often not very deep and contain minerals released into them by the weathering of the parent material quite near the surface. By contrast, in tropical latitudes soils are often geologically old, and because of a long period of development they have become deep. The parent material is too deep to supply minerals to the soil surface by weathering and the soils have become depleted of minerals which have been washed out by rainfall over millions of years.

Because of its huge area and global importance, let us now look in a little more detail at a tropical rainforest ecosystem. The Amazon basin in South America provides a good example.

3.2   *Tropical rainforest ecosystems of Amazonia*

The Amazon basin is a huge area, about the size of Australia, drained by rivers which carry one-fifth of the entire world's fresh water. The rainforest which covers the area is a biological treasure-house of species, where probably greater than 10% of the world's flora and fauna are to be found, and where, at a conservative estimate, the living trees and soil hold over 10% of the carbon in the terrestrial biosphere.

These enormous biological riches, particularly the tall rainforest trees and lush vegetation, have long impressed explorers and scientists with the seeming fertility of the area. Alfred Russell Wallace spent five years in Amazonia and, like nearly all visitors to the region until nearly the middle of the twentieth century, he mistook the luxuriance of the vegetation there to indicate great soil fertility:

> I fearlessly assert that here, the 'primaeval' forest can be converted
> into rich pasture and meadow land, into cultivated fields, gardens and
> orchards containing every variety of produce with half the labour and,
> what is of more importance, in less than half the time that would be
> required at home.   (A. R. Wallace, 1853, *Travels on the Amazon*)

As if to take up this challenge, the Brazilian Government in the 1970s began encouraging settlement and forest clearance for agriculture in Amazonia, most of which lies within their territory. The intention was to offer 'land without men for men without land' and to develop Amazonia economically. The construction of the trans-Amazon highway, begun in 1970, opened up Amazonia to colonisation by immigrants from Brazil's poor north-east region, who were attracted by the prospect of unlimited fertile land. Unfortunately, the fertility of the land was an illusion. The tragic error of believing that large trees indicate a fertile soil has led to an environmental threat of huge, and some fear global, proportions.

Refer back to Figure 7.8, which shows the distribution of nitrogen between the biomass and soil in a typical tropical rainforest.

Q   Where would you expect most of the nitrogen and other mineral
    elements to be found in a tropical forest ecosystem?

A   Above ground, in the trees themselves.

There is considerable variability of soils and forest types in the Amazon basin, but most of the soils are millions of years old and have had their nutrients washed out by rainfall. Even the most fertile Amazon soils lack the fertility and nutrient-retaining capacity of more recently formed soils in the temperate zone. The rainforest trees are able to grow on these soils because they are extremely efficient at recapturing nutrients released by the decomposition of organic matter on the surface of the soil. In some Amazonian forests there is a mat of fine tangled roots *above* the soil surface. Root tips attach themselves to fallen leaves, branches and fruits and withdraw nutrients from them before they have even reached the soil itself.

Clearing the tropical forest by burning returns most of the nutrients held in the vegetation to the soil, but the soil in many areas has little retentive capacity for these nutrients. Within a few seasons of growing crops soil nutrients are exhausted and the land has to be abandoned, or converted to very low-grade cattle pasture. The farmer then moves on to a new patch of forest, burns it, and in order to scratch a meagre living for a few more years, destroys more tropical forest. This process of shifting cultivation is sustainable when population density is low because abandoned land has time to recover before a farmer returns. However, in Amazonia, according to a conservative estimate, this method of cultivation by 1985 had already led to the disappearance of 10% of the rainforest. In 1987 it is estimated that 20 million hectares of forest were burnt. Some of this total represented areas that had been cleared before and had then been recolonised by trees, but 8 million hectares were virgin (primary) rainforest. All estimates of the rate of deforestation at that time agreed that the destruction was accelerating. Satellite images taken in 1988 showed between 6000 and 8000 fires burning in Amazonia on any one occasion.

Q   Recall from Chapter 5 that trees may locally influence the water cycle of an area. How do they do this?

A   Trees transfer water from the soil to the air by evapotranspiration.

It has been estimated that half the rainfall of the Amazon region originates locally from the recycling of water through transpiration by trees. Deforestation may therefore not only affect the soil and vegetation of the region but its climate too, since a 50% drop in the water returning to the atmosphere would reduce clouds and rainfall.

The study of nutrient cycling in the Amazonian rainforest ecosystem shows the folly of forest clearance there, but it also suggests how such fragile ecosystems can be exploited without destroying them.

Because of the ease with which nutrients are lost from Amazon soils following deforestation, any sustainable exploitation of them must make use of the nutrient-retaining capacity of the intact rainforest ecosystem itself. This may be done, for example, by harvesting natural products from the forest such as brazil nuts and rubber in the Amazon or rattan canes in Malaysia and Indonesia, by limiting shifting cultivation to small areas bordered by forest which will trap released nutrients as they run off in streams, or by adding organic matter to the soil to make it more retentive. In prehistoric times certain Indian tribes developed soils suitable for cultivation by adding organic matter. The patches of black, humus-rich soil that they produced have been found in some parts of the Amazon and are still fertile.

Not all tropical or subtropical soils are old and infertile. River deltas such as the Nile in Egypt are extremely fertile because they receive inputs of mineral-laden silt from upstream.

Q There are other tropical regions where the soils are young and fertile. What geological events mentioned in Chapter 5 give rise to these? Hint: Where 'new' rocks appear at the surface they provide a parent material rich in minerals.

A Volcanic eruptions, which are particularly common at the boundaries of tectonic plates, produce parent materials that give rise to fertile soils. The lava soils of Indonesia are an example.

3.3 Mineral elements and primary production

The role of decomposition in ecosystems demonstrates how the fate of primary production in an ecosystem may affect mineral cycling. The reverse may also happen when a shortage of minerals limits primary production. Nowhere is this more clearly demonstrated than in lands where an essential plant nutrient limits agricultural production.

In much of South Australia the soil is naturally deficient in nitrogen, phosphorus and sulphur. The native vegetation of the region is adapted to these mineral deficiencies, but grows slowly and provides poor fodder for sheep, so more productive plants were introduced, but these require the addition of fertiliser.

Most important of the introduced plants is subterranean clover *Trifolium subterraneum* (which, like the peanut, buries its seeds). Like all plants in the pea family (collectively called 'legumes'), this species has nodules on its roots; these contain bacteria which fix nitrogen from the air. When this species was first introduced, pastures sown with it were fertilised with superphosphate of lime, which supplied both phosphorus and sulphur. Although this treatment should have provided all the minerals that were required, the clover grew only poorly. Researchers later found that another essential chemical element was missing from the soils – molybdenum. This is required by plants in only minute amounts and is called a trace element. Addition of only 5 grams per hectare was enough to promote vigorous growth of clover when it was also fertilised with phosphorus and sulphur. The effect of the addition of molybdenum on primary productivity was so great that soils which formerly required 16 hectares to support 10 sheep could now support the same number on a single hectare.

Activity 4

Q1 Assuming that sheep ate the same amount of food per head before and after molybdenum was added to the fertiliser, what effect did the addition of 5 g per hectare of this element have upon energy flow in the ecosystem?

Q2 The addition of molybdenum to the South Australian pasture altered the cycling of nitrogen as well as the primary productivity of the ecosystem. How did this occur?

3.4 Soil erosion

Soil erosion is one of the most ancient environmental problems to have plagued humans, and with which humans have plagued the Earth. The ancient Greek philosopher Plato was familiar with deforestation and its effects on soil and water. Writing in the fourth century BC he imagined what the countryside around Athens might once have looked like:

> The soil was more fertile than that of any other country and so could maintain a large army exempt from the calls of agricultural labour . . . [Now] You are left with something rather like the skeleton of a body wasted by disease; the rich, soft soil has all run away leaving the land nothing but skin and bone. But in those days the damage had not taken place, the hills had high crests, the rocky plain of Phelleus was covered with rich soil, and the mountains were covered by thick woods, of which there are some traces today . . . The soil benefited from an annual rainfall which did not run to waste off the bare earth as it does today, but was absorbed in large quantities and stored in retentive layers of clay, so that what was drunk down by the higher regions flowed downwards into the valleys and appeared everywhere in a multitude of rivers and springs.   (Plato, *Critias*, Penguin Classics Edition, 1971, p. 132)

Plato was wont to combine myth and reality in his glorification of the Athenian past. Although his account of the deforestation around Athens is not to be taken literally, Plato showed an understanding of soil erosion which cannot have come entirely from his imagination.

Today, soil erosion is a world-wide problem that seriously threatens global food production. Topsoil forms at a rate of 0.3–2 tonnes per hectare per year, but a survey of erosion rates found that in many parts of the world it is currently being lost at rates that far exceed this. Erosion by water selectively carries away the smallest soil particles and the organic matter from the soil. These important soil ingredients help its water-retaining capacity. Once erosion has begun, the soil becomes progressively less absorbent to water, erosion channels form and the rate at which water runs off into rivers accelerates. Ironically, water shortage to plants is therefore the most serious consequence of erosion.

The problem of soil erosion is worst on mountain slopes that have been cleared of forest and newly taken into cultivation, but areas where cultivation has been practised for millennia are suffering too. In the basin of the Yellow River in central China the fertile soil is being carried away by water erosion at a rate of 100 tonnes per hectare per year. In fact this is an ancient problem in the area and the Yellow River owes its name to the burden of silt it carries. However, the problem is getting worse, and parts of the river have been raised 7 m above the surrounding plain by the deposition of silt.

Britain also has soil erosion problems. The Soil Survey of England and Wales has estimated that 44% of arable land is at risk. Upland areas of blanket bog are susceptible to peat erosion. Mosses make an important contribution to primary production and peat formation in blanket bogs, particularly several species of *Sphagnum* and a moss species called *Racomitrium lanuginosum*. Because mosses have no roots, they are particularly dependent upon rainfall for a nutrient supply and are sensitive to the chemical composition of the rain and atmosphere.

In the southern Pennines of England, around Manchester and other industrial towns, the blanket bogs are unusually lacking in the main peat-forming mosses, although the remains of these species can be identified at a depth of 50 mm or more in the peat. Soot makes an appearance near the surface of the peat profile at the depth where *Sphagnum* and *Racomitrium* disappear, suggesting an explanation for the scarcity of the mosses today. These moss species appear to be sensitive to atmospheric pollution by sulphur dioxide. It has been calculated that between the mid-nineteenth and mid-twentieth centuries this gas could have been present in sufficient concentration to severely retard the growth of *Sphagnum* and *Racomitrium*.

In the unpolluted habitat in which these mosses live, their ability to take up nitrogen compounds from the weak solution normally found in rainwater enables them to grow in a very nutrient-poor ecosystem. Unfortunately, the efficient nutrient-capturing mechanisms which are an advantage in the environment of an unpolluted blanket bog also make the mosses which grow there highly sensitive to atmospheric pollution.

The effect of atmospheric pollution on such a major component of the blanket-bog ecosystem slows or even halts the build-up of peat and can make polluted areas susceptible to severe erosion.

3.5   Summary

In terrestrial ecosystems decomposition occurs mostly in the soil, where a wealth of animals, fungi and bacteria are to be found. In temperate ecosystems earthworms are an important component of the soil fauna, and in the tropics termites play a very important part in decomposition. The activities of decomposers are influenced by temperature and moisture, which results in a broad, global correlation between climate and the distribution of minerals between the soil and the living biomass.

In tropical rainforest, the soils tend to have poor nutrient-retaining power, and most mineral nutrients are in the biomass. These may be quickly lost if the forest is felled and burnt. In some circumstances, such as in large areas of South Australia, mineral deficiencies can limit primary production.

Erosion is an ancient threat to the productivity of soils, and affects large areas of land all over the world. A significant proportion of Britain's arable farmland is at risk from soil erosion, and peat erosion threatens some upland bogs.

4   Marine ecosystems

4.1   The animals and plants

All marine organisms, including fish, may be divided into those which feed in the body of the sea (called the pelagic zone) and those which feed on the sea bed (the benthic zone). In the pelagic zone live the **plankton**, which is the general name given to all the organisms that are relatively poor swimmers and whose distribution is determined by ocean currents. The plankton include bacteria responsible for decomposition, plants (phytoplankton) and small animals (zooplankton). Most fish begin life living in the plankton (Figure 7.9).

Most fish, and all sea mammals, sea turtles, penguin and squid also inhabit the pelagic zone, but their swimming abilities place them in the assemblage called the nekton. The majority of the nekton are carnivores (Figure 7.10).

The sediment on the sea bed contains a whole community of invertebrate animals of its own (Figure 7.11). These animals live on the detritus which sinks down to them from above, and include snails and bivalve molluscs, polychaete worms, shrimps, sea cucumbers, anemones and brittle stars (a kind of starfish). Many of these animals live buried in the sediment itself.

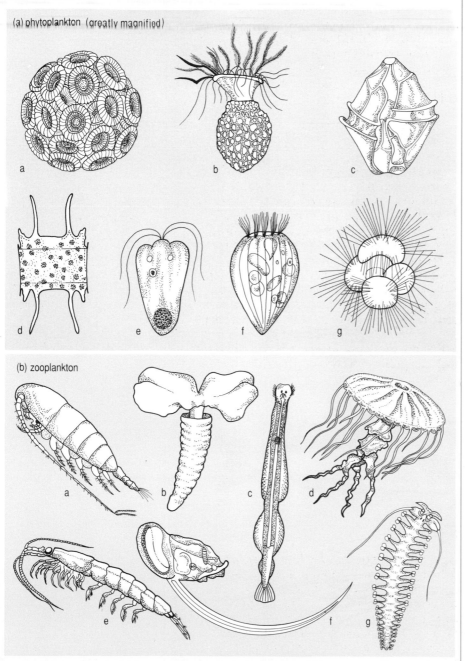

▲   *Figure 7.9   Examples of plankton. Key: (a) Phytoplankton: a, coccolithophore; b, tintinnid; d, dinoflagellate; d, diatom; e, flagellate; f, ciliate; g, foraminiferan. (b) Zooplankton: a, copepod; b, pteropod; c, chaetognath; d, jellyfish; e, euphausid (krill); f, appendicularian; h, polychaete.*

4.2   *Marine primary production*

In terrestrial ecosystems the most important primary producers are very
often trees or other relatively long-lived plants. The sheer size of trees and
the indigestibility of wood causes the energy and carbon contained in them
to be released only slowly into the ecosystem. By contrast, in marine

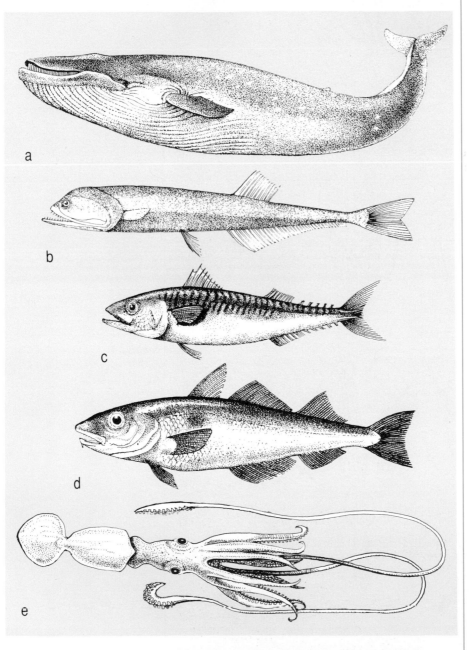

▲   *Figure 7.10   Examples of nekton. Key: a, blue whale* Balaenoptera musculus
(can reach 30 m in length); b, a deep sea fish Cyclothone elongata *(up to 100 mm);
c, mackerel (up to 350 mm); d, haddock (up to 900 mm); e, squid (body plus short
tentacles up to 300 mm).*

ecosystems the primary producers are mostly microscopic phytoplankton, some so tiny that they pass through nets of the finest mesh that are used by marine biologists to sample the ocean. These tiny plants have very brief lives, multiply very fast, and decompose very quickly when they die.

Q   What effect would you expect the difference between the major kinds of primary producers on land and in the sea to have upon the rate at which carbon cycles within the two kinds of ecosystem?

A   Carbon should cycle much faster in the sea because it is released from the primary producers very much more rapidly than on land.

As with all small organisms, the small size of phytoplankton is associated with a very high potential rate of population increase when conditions are right. This produces the sudden blooms of phytoplankton that are characteristic of both lakes and the sea. In lakes such blooms occur particularly when the waters are polluted by nutrients such as agricultural fertilisers that run into them from the surrounding land.

   Another difference between terrestrial and marine ecosystems is in the accessibility of minerals to the primary producers. As we have seen, the soil is the main source of minerals for terrestrial plants, but minerals contained in sediments on the sea bed are inaccessible to phytoplankton unless turbulence brings them to the surface. Phytoplankton are confined to the upper layer of the ocean, called the photic zone, where there is sufficient light for photosynthesis. In both lakes and oceans zooplankton such as copepods (see Figure 7.9), which feed upon phytoplankton, tend to migrate vertically between lower levels where they spend the day, and the photic zone at night. One hypothesis as to why they do this suggests that it protects them from predation by fish which find their food by sight.

   In deeper lakes and some parts of the sea, the water column is often stratified in summer, with a warm surface layer of water sitting like a lid over a colder layer of water beneath. This prevents nutrients being returned from lower levels to the photic zone, until winds or a change in air temperature cause an overturn which mixes the water.

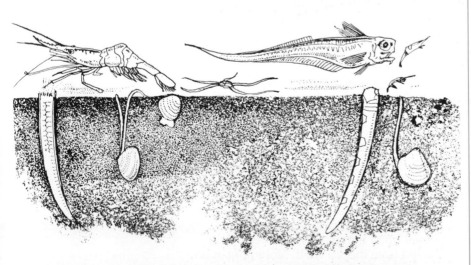

▲ Figure 7.11   Examples of the benthic fauna, including polychaete worms, bivalves, brittle star and crustaceans.

Stratification is not a problem in flowing water. Animal life in streams and small rivers often depends on the input of dead material from their banks and from upstream.

The restriction of phytoplankton to the photic zone has strong effects upon the distribution of primary productivity in the sea. Phytoplankton production in the open sea tends to be very patchy, as ephemeral pools of nutrients appear, are mopped up by phytoplankton blooms and are then dispersed by surface winds and currents.

Table 7.1 shows estimates of primary productivity for different marine ecosystems of the world. You should now do Activity 5, which will help you interpret the table.

Table 7.1

Ecosystem	Area/ (10^{12} m^2)	Net primary production/ (g per m^2 per year)	Total net primary production/ (10^{15} g per year)
Open ocean	332	230	76.2
Upwelling zones	0.4	918	0.4
Continental shelf	26.6	660	17.6
Attached algae and reefs	0.6	2500	1.6
Estuaries	1.4	1500	2.1
Total or average	361.0	271	94.0

Activity 5

Q1 In Table 7.1, which ecosystems have the highest and which the lowest primary productivity per m^2?

Q2 Which ecosystems have the highest and which the lowest primary productivity in total?

Q3 From the point of view of the global carbon cycle, which type of marine ecosystem is most important?

Q4 From the point of view of fisheries, which types of marine ecosystems do you think are the most important?

Estimates of primary productivity in the ocean are often inaccurate because conditions in the open sea are so variable and patchy, and because difficulties in sampling make it possible to underestimate the contribution made by microscopic plankton. Remote sensing of the oceans by satellite potentially offers a means of obtaining more accurate estimates, but because such measurements are based upon the colour of the ocean they have to be calibrated by samples taken from the sea itself. Notwithstanding the uncertainties in the values estimated in Table 7.1, the relative contributions of each kind of ecosystem to the whole are probably correct.

4.3   Summary

Marine organisms may be divided according to where they are found in the water. Those in the body of the sea are pelagic, and those on the sea bed are benthic. Plankton, which include phytoplankton and zooplankton, are small and live in the pelagic zone. Nekton are the larger marine animals, most of which are carnivores. On and in the sea bed there is a community of animals living upon detritus that sinks from above.

The greatest concentration of marine primary productivity is in shallow waters and upwelling areas. Although the open ocean does not have great productivity per unit area, it is so vast that it makes the biggest total contribution to ocean primary production.

5   Populations

5.1   Introduction

Energy flow and nutrient cycles are the consequences of complex underlying ecological processes. To understand these processes properly, we will need to look in greater detail at the organisms involved.

As a rule of thumb, the importance of a particular species in an ecosystem depends upon its *numbers*. A single winter moth is of little consequence, but an oaktree-full is a calamity. One earthworm per square metre will have little effect upon soil properties, but one hundred earthworms may transform the soil totally. A single Nile perch in Lake Victoria would not be noticed, but a population of these predators can cause the extinction of endemic cichlids.

Q   Though these statements are true if taken at face value, what is the important oversight in them about the significance of small numbers?

A   In favourable circumstances small numbers of organisms *can* be the founders of large populations, and a few individuals of very large species can make a significant impact on an ecosystem.

A **population** is a collection of organisms belonging to the same species, found within a limited geographical area. If they are sexual, this is the area within which individuals interbreed. All organisms have the potential to reproduce. The majority of species reproduce sexually, producing offspring which are to some degree genetically different from their parents and from each other. There are species however, such as dandelions, and aphids at certain stages of their life cycle, which are able to reproduce *asexually*. A single individual may therefore found a population. This is one reason why it is so difficult to eradicate aphids from the garden!

How the numbers of individuals in a population are determined is a fundamental question in ecology, and one which also underlies many practical issues: the conservation of rare species, the control of pests, forestry, the regulation of fisheries, and the control of disease. The principles of population ecology also apply to the human population, though we alone among species have the power to choose how we manage our own population.

5.2   *Birth, death, migration*

Increase is inherent in all populations. Whether or not this potential is realised in a particular place at a particular time depends upon the balance of four processes: birth, death, emigration from the population and immigration into the population. These four processes are described by **population parameters** that measure the **birth rate**, the **death rate** and the rates of immigration and emigration, as a proportion of the existing population.

All four population parameters vary with time. The size of a population at any particular moment will depend upon the relationship between the processes which add new individuals (birth and immigration) and those which cause a loss (death and emigration).

Q   Imagine a population of deer (say) with a birth rate of 200 calves per thousand adults per year and no net migration. If the death rate is 50 per thousand adults per year, would the population increase, decrease, or stay constant in numbers?

A   The population would increase each year by the difference between the birth rate and the death rate: i.e. by 200–50 = 150 individuals per thousand adults per year. That is 150/1000, which is a rate of population increase of 15% per year.

Q   If the difference between the birth rate and death rate remained constant at a value of 15% per year, how would you expect a population of a thousand deer to increase in three years?

A   At the end of the first year there would be a net increase to 1000×(115/100) = 1150. At the end of the second year this would have increased to 1150×(115/100) = 1323, and 1323×(115/100) = 1521 at the end of the third.

This kind of multiplication by a fixed percentage each year is mathematically the same as the calculation of compound interest. It is said to produce a *geometric* increase in population size. By the tenth year the population will have more than quadrupled, as shown in Figure 7.12.

As Figure 7.12 shows, a constant rate of population increase will result in an *accelerating* increase in the total size of the population. This operates just like the positive feedback process we discussed in Chapter 5 because every change produces more change in the same direction. If the death rate exceeds the sum of the birth rate and immigration rate, then this will lead to an accelerating *fall* in population size which can rapidly result in the extinction of the population if it continues unchecked.

In statistical terms, the likelihood of a female giving birth changes with her age; likewise, the risk of any individual dying is also age related. Figure 7.13 illustrates this in a classic study of red deer. Actual birth rates and death rates for a population will therefore depend upon the proportion of individuals in those age groups which produce the most young, or the proportion that are old and are at greatest risk of death.

The spread of individuals through all the age groups from newborns to oldsters is called the **age structure** of the population. The red deer is given as an example in Figure 7.14. An increasing population characteristically has an age structure with a high proportion of youngsters, whereas a population with a high proportion of oldsters is likely to wane in numbers.

In practice, of course, populations cannot increase or decrease indefinitely, so something must keep population size within bounds. What is responsible for this?

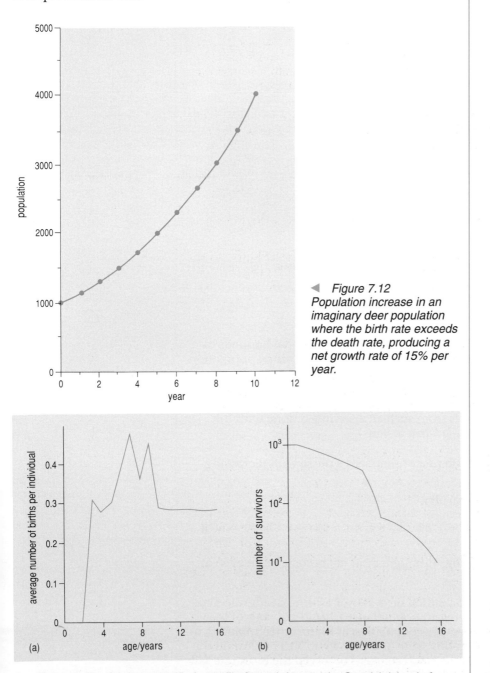

◄   *Figure 7.12*
Population increase in an imaginary deer population where the birth rate exceeds the death rate, producing a net growth rate of 15% per year.

▲   *Figure 7.13   (a) Age-specific fecundity for red deer on the Scottish island of Rhum. (b) Age-specific survival of the red deer. Note the scale in powers of ten.*

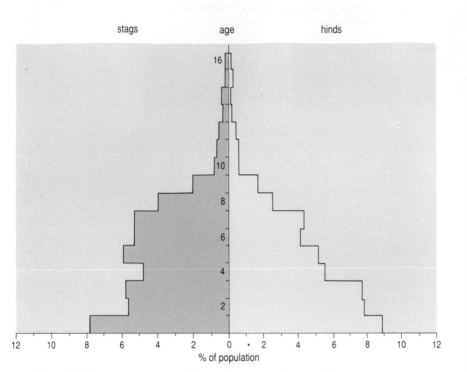

stags   age   hinds

▲   Figure 7.14   Age structure of the red deer population of Rhum in 1957.

5.3   What limits populations?

The answer to this question can be seen from observing what happens to
birth rates, death rates, emigration and immigration rates when population
density rises. A very simple experiment to demonstrate this, which anyone
can perform in a few plastic cups on their window sill or in their garden, is
to sow seeds of a quick-growing plant such as cress at a range of densities.
Figure 7.15 shows some results from such an experiment done with the
weed shepherd's purse (*Capsella bursa-pastoris*).

Q   What happened to the birth rate of shepherd's purse with increasing
    density?

A   The birth rate fell very sharply.

Q   What happened to the death rate of shepherd's purse with increasing
    density?

A   The death rate rose steadily, although it did not become large even at
    the highest density.

The relative contribution that a falling birth rate or a rising death rate make
to limiting numbers varies between populations. Changes in both of these
parameters operated in the red deer population of Rhum when numbers in
one part of the island rose steadily from 19.8 per km² to 30.5 per km² in the
space of ten years. During this period ecologists observed the birth rate fall
by half (Figure 7.16a) and the mortality of calves in winter rise from less
than 5% to more than 20% (Figure 7.16b).

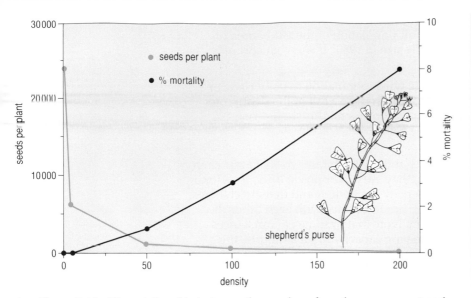

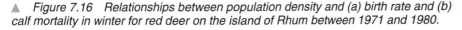

▲ Figure 7.15 The relationship between the number of seeds sown per pot and birth rate (seeds produced per plant) and death rate of seedlings of shepherd's purse sown at a range of five densities.

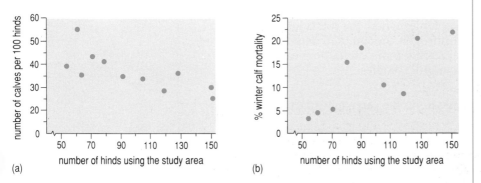

▲ Figure 7.16 Relationships between population density and (a) birth rate and (b) calf mortality in winter for red deer on the island of Rhum between 1971 and 1980.

Q What effect would you expect these changes in birth rate and death rate to have upon the further increase of the red deer population?

A A reduction in the birth rate and an increase in the death rate should check the further increase of the red deer population.

When the numerical value of population parameters changes with density, such as occurred in the red deer population of Rhum, or the pot experiment with shepherd's purse, the parameters are called **density-dependent factors**. The special significance of density-dependent factors lies in the fact that *only* parameters which alter in a density-dependent way can stabilise population size. Density-dependent factors impose a ceiling on the rise of population numbers. If this ceiling is determined by the availability of food or space it is called the carrying capacity of the habitat for the species in question.

Q   Would you expect the density-dependent death rate of red deer calves
    to have changed if in 1981 the adult population of the study area on
    Rhum had been reduced to half its size in 1980? If so, what sort of
    change would you expect?

A   Density-dependent factors change when population density rises, or
    when it falls. A fall in density should produce a fall in the death rate.
    Since the birth rate was also density dependent, this should rise as
    density falls.

Remember that because the death rate and the birth rate work against each
other in determining population size, they vary in opposite directions (one
up, the other down) when they are density dependent and density changes.
When the death rate and the birth rate are in balance with each other the
population is said to be at equilibrium. Note that in some circumstances
this equilibrium can fall *below* the carrying capacity if, for example, the
death rate is strongly influenced by a predator.
    What is the cause of the change in the death rate and the birth rate
when density changes? In the case of the red deer of Rhum, the answer
seems to be a shortage of their plant food, which consists mainly of grasses.
The growth of these grasses is affected by the weather, and their nutritional
value for the deer is affected by soil fertility. Both of these environmental
factors have been shown to affect the deer through their food supply. On
Rhum, grass growth is favoured by high rainfall between May and July and
low rainfall in September.
    Grasses growing near gull colonies where the soil is fertilised by the
birds' droppings have a higher nitrogen content. Red deer hinds observed
grazing in these areas had significantly more calves per year than those
females which did not graze near gull colonies.

Q   What effect do the gulls have upon the carrying capacity of the local
    area for red deer?

A   The gull's droppings improve the quality of the deer's food supply by
    fertilising the grass. This raises the carrying capacity of the deer's
    habitat locally.

Red deer populations elsewhere in Scotland are also regulated by the
availability of their food, although the details of their ecology are different.
On the Glenfeshie estate in Inverness-shire for example, the red deer's diet
is mainly heather. Heather growth at this site is greater in warm, dry
summers than in cool, wet ones, and consequently there is more winter
food for the deer following the former kind of summer than the latter. The
number of red deer hinds on the estate in spring is closely related to the
amount of heather production from the previous year, which was available
to them as winter food (Figure 7.17).
    In contrast to the Rhum population, the variation in hind numbers with
food supply shown in Figure 7.17 was not due to hind mortality at
Glenfeshie during winters when heather was scarce. Instead, the ecologists
studying this population suggested that the variation in the number of
hinds from year to year was due to animals leaving the area when little
food was available. Here, then, we have a case where emigration may have
a significant impact upon the size of the local red deer population.
    It should be noted that the details of how a population is regulated may
change from place to place, and even from year to year. What does not

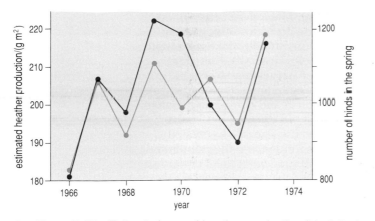

▲   *Figure 7.17   Estimated annual heather production (black line) and the number of red deer hinds counted (green line) at Glenfeshie, Inverness-shire in 1966–73.*

change is the crucial role of density dependence, whether it acts through death rate, birth rate, emigration or immigration rates.

Rates of immigration and emigration are a particularly important factor in determining population size in highly mobile animals, especially when their food supply is very variable. Crossbills (*Loxia* spp.) are birds which feed upon the seeds of coniferous trees, whose availability varies considerably from year to year. When seeds are in short supply, large migrations of these birds occur across Europe.

Among butterflies the red admiral (*Vanessa atalanta*) and the painted lady (*Cynthia cardui*) breed, but rarely survive over winter in Britain. The adult butterflies found at the beginning of the season here are all immigrants whose offspring return to southern Europe. The size of populations of these species in Britain may be determined by events at the other end of their migration, rather than by their predators or food supply here.

5.4   *Harvesting from wild populations*

Although most of our food comes from farmed populations of animals and plants whose rates of growth, reproduction and death are manipulated to supply us with a regular harvest, many wild or semi-wild populations are harvested too. Numerically the most important among these are marine populations of fish, squid, shellfish and (controversially) marine mammals such as seals and whales. Since we don't directly control the birth rates of these animals, there is the possibility that removing individuals from some of these populations might increase the death rate above the point where it is balanced by the natural birth rate. Removing females could also lower the birth rate, making a balance more difficult to achieve. If this were to happen, then the harvested population would start to decline.

Of course there is a limit to how long a population can continue to provide a harvest if fishing or some other hunting activity is causing it to decline. We saw in Chapter 6 that some species have been hunted to extinction. The objective, then, of a rational exploitation of wild populations should be to maintain a **sustainable harvest**. In theory it should be possible to remove some fish, say, from a population where the birth rate exceeds the natural death rate without risking extinction.

Furthermore, if the natural death rate or the birth rate are density dependent, losses caused by fishing will be made up by a compensatory adjustment in these rates. Either of these situations should make it possible to sustain a harvest from a population indefinitely. However, there are some fundamental practical problems with these harvesting strategies.

There are always limits to how far density dependence can compensate, and these limits must be known. Also, natural birth and death rates change, as for example when *El Niño* occurs off the coast of Peru and causes the anchovy population to crash (Chapter 5). Unless the size of a harvested population is known accurately, and the factors influencing the population parameters are understood, it is difficult to determine what level of harvest is sustainable.

5.5   Human demography

The study of trends in the size and composition of the human population is called demography. Although this term originally referred exclusively to the study of the human species, it is now used for the study of populations of all animals and plants. Demographers of the human population like to joke that theirs is 'the study of people broken down by age and sex'. Classifying the members of a population by their age and sex is a tool that is useful for describing all populations (see Figure 7.14, for example).

Q   Why does age structure, broken down into males and females, provide a useful picture of a population?

A   Because birth rates and death rates vary with the age and sex of individuals. This information is required to predict how the population will change in the near future.

History of human population

Accurate censuses of the human population are essential to planning, but are difficult to achieve in many countries, even today. The size of the human population of the world is therefore known only approximately. It is estimated that it surpassed 5×10^9 (5 billion) in 1987 and is projected to reach 6.3×10^9 (6.3 billion) by the year 2000. What is most alarming about this

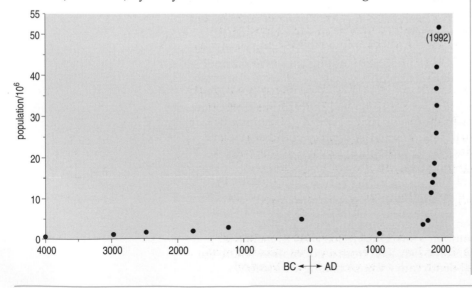

◀ *Figure 7.18*
Population growth in Egypt from 4000 BC to the present.

figure is that it is growing at about 2% a year, which means that it will double in only 35 years. Historical records of population show this growth in perspective. For example archaeologists have been able to reconstruct the history of population growth in Egypt from 4000 BC to the present (Figure 7.18). The reconstructed trend of world population follows the same pattern (Figure 7.19).

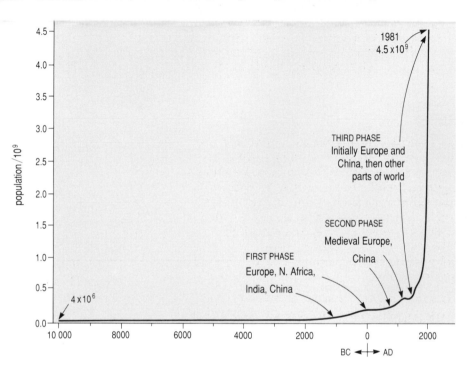

◄ Figure 7.19
Global population growth from 10 000 BC to the present; (the trend continues upwards).

Present patterns of population change

Global averages hide a picture of human population growth that is more complicated when the statistics are broken down region by region. Table 7.2 shows that population growth is slight in North America, Europe and the USSR and highest of all in Africa. Note that Africa has the highest growth rate of all the regions shown because it has the biggest difference between birth rate and the death rate, and that both of these rates are high.

Q   Africa has a higher *rate* of population growth than China, but which of the two areas would have had the greater increase in *absolute numbers* of people in 1992, according to the information in Table 7.2?

A   In 1981 Africa's population was 682×10^6 and China's was 1188×10^6. To find the absolute increase in population for each of these areas multiply the rate of population growth by the total population:

Africa: $(682 \times 10^6) \times \dfrac{3}{100} = 20.46 \times 10^6$ China: $(1188 \times 10^6) \times \dfrac{1.5}{100} = 17.82 \times 10^6$

Thus the total population increase in Africa, as well as the growth rate, exceeds that in China. (In the first edition of this book, this calculation was done for the 1981 data, when the growth rates were similar, but both

Table 7.2   Population statistics for various regions and countries

	Total population/10^6	Density/ per km^2	Births/ per thousand	Deaths/ per thousand	Annual rate of increase (%)
World	5479	40	27	10	1.7
Tropical regions					
Central America	119	48	32	6	2.3
South America	3044	17	27	8	1.9
Caribbean islands	35	147	25	8	1.5
Africa	682	22	45	15	3
South-east Asia	416	103	30	9	2
Melanesia, Micronesia and Polynesia	6.5	239	32	7	2.1
Temperate regions and countries					
North America	283	13	16	9	1.0
Europe	512	100	13	10	0.4
former USSR	285	13	19	10	0.8
China	1188	124	–	–	1.5

Rates are annual averages for the period 1985–92; total population and density are for 1992.
Source: United Nations (1992) *Demographic Yearbook*, United Nations, New York.

populations were correspondingly smaller. At that time (1990), the Chinese population increase was greater than that of Africa. This change shows how rapidly the greater growth rate can outweigh the greater population.) If the growth rates do not change, the African population will exceed that of China in a few decades.

A century or more ago, the birth rates and death rates of the regions which now have low population growth rates were as high as those of third world countries today. The transition from high rates of birth and death to present levels took place in parallel with the industrialisation of the developed countries, but the two rates did not adjust simultaneously. The death rate fell first, followed by a fall in the birth rate, and the lag between the two produced a concurrent period of rapid population growth.

In demographic terms the present high rates of population growth in developing countries are also the result of an improvement in the death rate, while birth rates remain high, or are even rising. This raises the obvious question, where will the food come from to feed the extra mouths?

When it occurred in their populations the countries of the first world coped with this problem by economic development. This was greatly aided by, and some would say based upon, colonial expansion and the exploitation of the resources of the undeveloped world. Economic development and birth control have both been advocated as solutions to the problem of population growth in the third world today.

The imbalance between a rising population and supposedly finite resources to feed and clothe the population was a problem addressed by an English clergyman by the name of Thomas Malthus at the close of the eighteenth century. He pointed out that all populations possess the inherent capacity to increase at a geometric rate (see Figure 7.12), but that food resources cannot be multiplied in this way. While Malthus was correct about the biological potential of populations to increase, his view of the limitations that control food production was simplistic.

The fact that a large proportion of the world's population does not receive enough to eat each day, or that there are perhaps 20 million deaths a year from starvation, would seem to support Malthus' case that the human population is limited by its food supply. To examine this argument properly we must ask *why* people starve, and not assume that it is a natural outcome of population growth or of 'over population'. After all, there is now no shortage of food in western Europe or North America, and indeed both the EU and the USA subsidise farmers' incomes to prevent them producing *too much* food. Wholesale famine has been abolished in India and China, two countries with fast-rising populations that have known starvation in previous periods when their populations were much lower.

To explore this argument fully would take up too much space here. The availability and exploitation of resources is a subject to which we will return, but for the moment bear in mind that there is an important difference between the human population and that of other species. Whereas other organisms are dependent upon a supply of food that has natural limits, resources used by the human population can be increased, and potentially fairly shared, by our ability to control the environment. Therein lies our hope and our responsiblity.

Although the limits of resources available to the human population are elastic and can be extended by technological means, there is usually an environmental cost in doing this. For example, increasing the intensity of farming, building more cities, and increasing the fish catch may all have environmental repercussions. Many of these expanding human activities have one important consequence in common: they erode the habitats available to wild animals and plants and put populations of these species in jeopardy. We have already looked at some aspects of this in Chapter 6. In the final section of this chapter we will take a brief look at the need to preserve genetic diversity within populations.

5.6   Summary

The numbers of an organism are a rough guide to its ecological importance. A population is a collection of organisms belonging to the same species, found within a limited geographical area. Geometric increase is inherent in all populations, but the *actual* size of a population depends upon the balance between four processes: birth, death, emigration and immigration. For individuals, the risk of death and the probability of giving birth both vary with age. The age structure of a population influences its future increase or decrease.

Density-dependent changes in birth rates and death rates can stabilise population numbers. When the stable population size is set by the availability of some resource such as food, it is called the carrying capacity.

In wild populations that are hunted, a sustainable harvest is one that does not cause the population to enter a continuous decline. In practice it is very difficult to determine the limits of a sustainable harvest.

The human population is subject to the same basic population processes as other animals and plants, though we uniquely have control over our birth rate and over the resources available to us. It is estimated that the human population surpassed 5×10^9 in 1987, and that it is growing at a rate of 2% every year. The expansion of the human population threatens other animals and plants.

6 *Genetic variation*

6.1 *Introduction*

In Chapter 6 we looked at taxonomic classification, and glimpsed some of the diversity of life to be found among living species. A further level of diversity is to be found *within* a species, in the genetic differences between its individual members. In sexually reproducing species, which include most animals and plants, every individual is genetically different from every other, with the uncommon exception of identical twins. There are inherited differences between individuals for a huge range of characteristics. A good example of genetic differences within our own species is to be found in blood groups. We inherit our blood group from our parents, but because both mother and father contribute a share to this inheritance, it is possible for a son or daughter to have a blood group different from either of the parents. The biochemical characteristics of an individual which cause organ transplants from others to be rejected are another example of inherited variation.

Genetic differences between individuals within a population create **genetic diversity** which is valuable in several ways. Two of the most important are in conferring disease resistance to populations, and in providing a source of characteristics that may be incorporated into crops by cross-breeding with their wild relatives.

6.2 *Genetic diversity and disease resistance*

Individuals of many species differ in their susceptibility to different diseases. This particular aspect of genetic diversity in a population has a positive value because it prevents all individuals being equally susceptible to all diseases. We can see the danger that this would present by looking at populations in which genetic diversity has been artificially reduced. Genetic uniformity is generally sought by plant breeders for new crop varieties, because it means that the seeds they sell farmers all reliably give the same results. In the 1960s maize farmers in the USA were nearly all growing the same popular variety, which turned out to be susceptible to a hitherto rare disease called southern corn leaf blight. In 1970 this disease came near to causing a nationwide crop failure of unprecedented proportions.

Genetic diversity can be lost from a population by repeated breeding among close relatives (called inbreeding). This tends to happen in the breeding of domesticated plants and animals during the course of the production of new varieties that have desirable characteristics (which could include resistance to some common disease). It may also happen in wild populations following a drastic reduction in population size, which leaves only a small group of closely related individuals to breed with each other. At some point in the past, perhaps during a geologically recent period of climatic change, the African cheetah appears to have gone through a reduction in population size which has left present day populations with very little genetic diversity. In some populations there

3al

has been so much inbreeding that skin grafts can be surgically exchanged between individuals without rejection. The low genetic diversity of cheetah populations was probably a contributory cause of susceptibility to a virus disease which killed 50–60% of animals in some populations.

6.3   Genetic diversity as a resource

The world's food supply depends upon a relative handful of plant species. Study Figure 7.20 and then consider the questions in Activity 6.

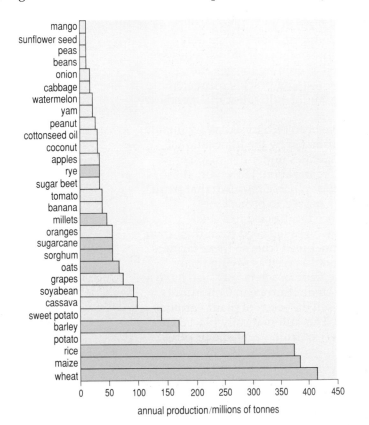

annual production/millions of tonnes

◄ Figure 7.20
The annual harvest of the 30 most important food crops world-wide. Crops that belong to the grass family are coloured green.

Activity 6

Q1 The total harvest of the crops shown in Figure 7.20 is about 2.7×10^9 tonnes. Approximately what percentage of this total is made up by the top four crops?

Q2 What proportion of the crop species listed in Figure 7.20 are grasses, and approximately what contribution do they make to the total annual production of food crops?

Because there is so little diversity *among* our major food crops, we must look to the preservation of the genetic diversity *within* each cultivated species to ensure and improve the security of supply.

Genetic improvement of crops requires a source of related plants with novel inherited characteristics that can be introduced into the crop by cross-breeding or by genetic engineering. Such sources tend to be geographically concentrated in centres of genetic diversity where wild populations of the crop's ancestors grow, often as weedy companions in the field alongside domesticated varieties. For example, the Near East is such an area for many types of wheat, Mexico for maize, the Andes for potato and Indonesia for the mango. The genetic diversity of plants in these centres is due not only to the presence of wild populations but also to the many different domesticated varieties that are cultivated by traditional farmers. The genetic diversity of crops grown by farmers who buy their seeds from a seed merchant is much lower by comparison, so a change from traditional to modern farming practices entails a loss of genetic diversity, although there may well be an increase in yield. The cruel irony is that the very success of plant breeders in producing better crops may undermine future crop improvement if a large number of old varieties are entirely replaced by a few new ones. As well as eroding the genetic diversity of the crop, the spread of intensive farming practices threatens wild populations of crop ancestors as more land is cultivated and weeds are more rigorously controlled.

The genetic diversity of crop plants have been recognised as a valuable resource and efforts to conserve it are co-ordinated by the International Board for Plant Genetic Resources, which operates under the auspices of the United Nations Food and Agriculture Organisation. However, at the same time as efforts are made to conserve the valuable resource that plant genetic diversity represents, laws have been passed to protect the proprietary rights of plant breeders. In France this has led to a seed company successfully prosecuting farmers who saved seed from their own crops to resow it. Resown seed often does not breed true, so these farmers' action would tend to increase genetic diversity as well as saving them money. Other laws designed to regulate the trade in seeds may also have an adverse effect upon crop genetic diversity. Under European Community regulations it has been illegal since 1981 to sell the seeds of plant varieties that are not listed in a 'common catalogue'. This automatically placed in jeopardy any rare or locally produced variety of fruit, vegetable or cereal that was unrecognised by officialdom.

6.4 Summary

All individuals (except identical twins) of species which reproduce sexually differ from one another in some of their inherited characteristics. These differences confer genetic diversity upon populations. An important consequence of this is that large populations can usually survive epidemic diseases because at least some individuals are likely to be resistant. Populations in which genetic diversity has been reduced by inbreeding (field crops, for example) tend to be vulnerable.

A large proportion of the world's food supply comes from a handful of plant species. It is important to preserve the genetic diversity of these crops as a source of future breeding material for crop improvement. This genetic diversity tends to be concentrated in a few geographical areas which often coincide with where the crops were originally domesticated from wild ancestors. Changes in agricultural practice and the increasing replacement of many traditional varieties by a few high-yielding ones may threaten the resource of crop genetic diversity. This may also be put in jeopardy by ill-conceived laws designed to regulate the trade in seeds.

7   Conclusion

The ecosystem and the population are somewhat abstract concepts, but they are ecological ideas which help us to get to grips with important processes. As we have seen, both are essential in understanding how the biosphere works, and in particular how parts of it may be manipulated for our benefit or inadvertently altered at our peril.

Answers to Activities

Activity 1

A completed version of Figure 7.1 is shown in Figure 7.21.

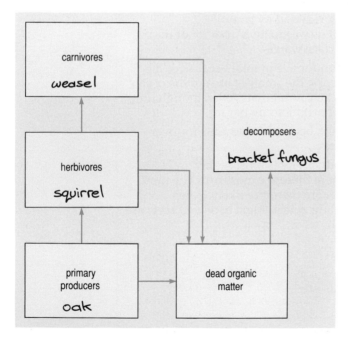

▲ *Figure 7.21   Compartments of a 'model' ecosystem, showing the pathways of energy flow (Figure 7.1 completed).*

Activity 2

Q1   3/324 = 0.009, or only about 1%.

Q2   The soil fauna consumes 131 of the 324 total annual net primary production, which is 131/324 = 0.40, or 40%. The soil fauna is therefore 40 times more important than the sheep and other herbivores in consuming primary production.

Q3   Most of it (278/324 = 86%) is decomposed by soil micro-organisms (bacteria and fungi) and is lost as heat in the process.

Activity 3

The proportion of the total nitrogen present in the soil decreases, from tundra where nearly all the nitrogen is in the soil, to rainforest where the soil contains only a small fraction of the total.

Activity 4

Q1   Since the area of pasture (food) required to support ten sheep dropped from 16 ha to 1 ha, there must have been a corresponding, 16-fold increase in the energy flowing through the ecosystem.

Q2   The molybdenum caused vigorous growth of subterranean clover which fixes atmospheric nitrogen in its root nodules. This provided a new input to the nitrogen cycle of the ecosystem.

Activity 5

Q1   Algal beds and coral reefs have the highest and the open ocean has the lowest primary productivity per m^2. Algal beds and coral reefs occur in shallow water where nutrients are tightly recycled by complex communities of animals and plants. The low productivity per m^2 of the open ocean is due to its nutrient-poor surface waters.

Q2   The open ocean is by far the most productive in total because of the sheer area it covers. Upwelling zones occupy the smallest area of the ocean and therefore contribute least to total ocean productivity, even though they exceed continental shelves on the basis of productivity per m^2.

Q3   The open ocean is the most important, because it accounts for over 75% of the carbon fixed in the sea in a year.

Q4   The ecosystems where primary productivity is most highly concentrated are likely to be the ones where there are most fish. The most important marine ecosystems for fisheries are therefore likely to be estuaries, continental shelves and upwelling zones. Algal beds and reefs are productive, but physically difficult to fish.

Activity 6

Q1   The top four crops and their harvests are:
       Wheat = 412×10^6
       Maize = 381×10^6
       Rice = 369×10^6
       Potato = 283×10^6
       Total = 1445×10^6 (which is the same as 1.445×10^9)
The total harvest of the crops shown in Figure 7.20 is about 2.7×10^9 tonnes. Therefore the top four crops comprise $1.445/2.7 \times 100 = 54\%$, or just over half of the total harvest.

Q2   Thirty crops are listed in Figure 7.20. Nine, or nearly a third of these, are grasses, and between them they account for more than half of the total harvest of all crops.

Further reading

BEGON, M, HARPER, J. L,. TOWNSEND, C. R. (1986) *Ecology*, Blackwell Science, Oxford.

KREBS, J. (1994) *Ecology: the experimental analysis of distribution and abundance*, 4th edition, Harper & Row, New York.

Acknowledgements

Grateful acknowledgement is made to the following sources for permission to reproduce material in this book:

Covers
*Front cover, clockwise from top right:*Val Corbett; Neil Leifer/London Features International Ltd; Roy Lawrance; Janice Robertson; Irene Ridge; British Nuclear Fuels plc; Janice Robertson; Sturrock/Network Photographers; Jonathan Silvertown; *centre*: Val Corbett; *back cover*: Information Service of the European Communities.

Colour plate section
Plate 1: Kenneth Scowen; *Plate 2*: Kenneth Scowen; *Plate 3*: European Space Agency/Science Photo Library; *Plate 4*: John Wightman/ARDEA; *Plate 5*: Canadian High Commission/ Photo Library; *Plate 6*: Heather Angel; *Plate 7*: Pat Morris/ARDEA; *Plate 8*: Heather Angel; *Plate 9*: Heather Angel; *Plate 10*: Jonathan Silvertown; *Plate 11*: Nature Conservancy Council; *Plate 12*: Nicolas Poussin, *Summer*, Louvre/Copyright Photo: Réunion des Musées Nationaux Documentation Photographique; *Plate 13*: François Gohier/ARDEA; *Plate 14*: Joseph Wright of Derby, *An eruption of Vesuvius, seen from Portici* (1774/76), reproduced by kind permission of the University College of Wales, Aberystwyth/Photo courtesy of The Tate Gallery; *Plate 15*: J.C. Ibbetson, *Ullswater from Gowbarrow Park*, Christie's Colour Library/Private Collection; *Plate 16*: J.M.W. Turner, *Lake Keswick*, reproduced by courtesy of the Trustees of the British Museum.

Figures
Figure 1.1: Ordnance Survey map of Cumbria, © Crown Copyright. Reproduced with the permission of the Controller of Her Majesty's Stationery Office; *Figure 1.2:* Toothill, J. (1988), 'The International Stage', photograph by Mike Williams/Pam Grant, *Lake District Guardian*, Lake District National Park Authority; *Figure 1.3:* Winpenny, D. (1987), 'Fell farmer trusts his homing instincts', *National Parks Today*, Countryside Commission, article and photographs by David Winpenny; *Figure 1.4:* 'Chapel Stile cited as typical ghost village', *Westmorland Gazette* © 6 March 1981; *Figure 1.5:* 'Langdale – Europe's most sought-after time ownership', Langdale Leisure Limited; *Figure 1.6:* 'The National Trust work at the Lake District' by Lord Shuttleworth, The National Trust Lake District Appeal, PO Box 3, Ambleside, Cumbria; *Figure 1.7:* Crook, T. (1973), 'High speed carve-up of Lakeland's park', *The Sunday Times* 7 January 1973, © Times Newspapers Limited; *Figure 1.8:* 'Monsters claim alarmist', *BNFL News* November 1988, British Nuclear Fuels plc; *Figure 1.9:* 'Cumbria solid geology', *Lake District National Park: Facts and Figures*, Nature Conservancy Council; *Figure 1.10:* Strahler, A. N. (1966), *Introduction to Physical Geography* 4th edition, John Wiley & Sons, © 1966 by Arthur N. Strahler; *Figure 1.11:* Adapted figure 19–21 from *Foundations of Chemistry* by Ernest R. Toon, George L. Ellis, & Jacob Brodkin, copyright © 1968 by Saunders College Publishing, a division of Holt, Rinehart & Winston, Inc., reprinted by permission of the publisher; *Figures 1.13 and 1.14*: NRPB Report 191, National Radiological Protection Board; *Figure 1.14*: also includes data from Caulfield, J. J. & Ledgerwood, F. K. (1989), *Terrestrial gamma ray dose rates out of doors in Northern Ireland*, Environmental Monitoring Report 2, Environmental Protection Division of the Department of the Environment (Northern Ireland); *Figure 1.15:* BNFL Annual Report 1987, British Nuclear Fuels plc.; *Figure 2.1:* Bennett, J. W. (1976), *The Ecological Transition*, Pergamon Press PLC, copyright © 1976 John Bennett; *Figure 2.2:* Hayden, B. (1981), 'Hunter/gatherers of the world, related to the Koppen-Geiger system of climate classifications' in Harding, R. S. O & Teleki, G. (1981), *Omnivorous Primates*, Columbia University Press, reproduced by permission of the authors; *Figure 2.3:* Martin, P. S. (1973), 'The Discovery of America', *Science* **173**, p.972, © 1973 by The American Association for the Advancement of Science; *Figure 2.5:* Bayliss-Smith, T. (1982), *The Ecology of Agricultural Systems*, p.47, Cambridge University Press; *Figure 2.7:* Quézel, P. *et al.* (1981), *Mediterranean-Type Shrublands* Vol. II, p.108, Elsevier B. V.; *Figure 2.10:* Trebilcock, C. (1981), *Industrialization of the Continental Powers 1780–1914*, Longman Group UK Ltd.; *Figure 3.1:* 'The mariners and the albatross', illustration by Gustave

Doré for the 1876 edition of S.T. Coleridge's *The Rime of the Ancient Mariner*; *Figure 3.2:* Levens Heritage, Levens Hall; *Figure 3.3:* Greenpeace/Weyler; *Figure 3.4:* Press Association; *Figure 3.5:* The Ecologist; *Figure 4.1:* Drawing by Vazquez de Sola, *Le Monde Diplomatique*, June 1992; *Figure 4.2:* 'Farce in Rio', *The Sun*, 15 June 1992, Rex Features Ltd; *Figure 4.3:* Pryor, G. (1992), 'And a good time was had by all', *Canberra Times*, 15 June 1992; *Figure 4.4:* 'Travel section' (Asides) Reprinted with permission of *The Wall Street Journal* © 1992 Dow Jones & Company, Inc. All rights reserved.; *Figure 4.5:* 'Growth and Greenery', *The Times*, 15 June 1992 © Times Newspapers Limited, 1992; *Figure 4.6:* 'Rio's sound and fury', *Japan Times*, 15 June 1992, © The Japan Times Limited 1992; *Figure 4.7:* 'Earth summit was just a beginning', *Bangkok Post*, 15 June 1992, © Post Publishing Company; *Figure 4.8:* D'vora Ben Shaul (1992), 'Rio 'Newspeak'', *The Jerusalem Post*, 15 June 1992, © The Jerusalem Post; *Figure 4.9: Osservatore Romano* (English edition), 15 June 1992, © Osservatore Romano;*Figure 5.10:* Hedgpeth (1957), Geological Society of America; *Figure 5.11:* Strahler, A. N. (1971), *The Earth Sciences* 2nd edition, p.235, Harper & Row, Publishers Inc. © 1963, 1971 by Arthur N. Strahler; *Figure 5.14:* Gribbin, J. (1988), 'The Greenhouse Effect', *New Scientist*, 22 October 1988, p.3, World Press Network Ltd; *Figure 5.15:* Barnola, J. M., Raynaud, D., Korotkevich, Y. S. & Corious, C. (1987), 'Vostock ice core provides 160,000 year record of atmospheric CO_2', *Nature*, **329**, p.410, © 1987 Macmillan Magazines Ltd.; *Figure 6.8:* Reprinted from *Trends in Ecology and Evolution*, **4**(2), Miller, D. J., 'Introductions and extinction of fish in the African great lakes', pp.56–59, Copyright 1989, with kind permission from Elsevier Trends Journals; *Figure 6.9:* Adapted from Rackham, O. (1986), *History of the Countryside*, p.70, J. M. Dent & Sons Ltd, © 1986 Oliver Rackham; *Figure 6.12:* Webb, N. (1986), *Heathland* p.44, Collins Publishers; *Figure 7.6:* Russell, E. J. (1961), *The World of the Soil*, p.145, Collins Publishers; *Figure 7.9:* Reprinted with permission from Parsons, T. R. & Tahalashi, M. (1973), *Biological Oceanographic Processes*, Pergamon Press PLC; *Figure 7.13(a):* Begon, M. & Mortimer, M. (1986), *Population Ecology*, Blackwell Scientific Publications Ltd; *Figure 7.13 (b):* Lowe, V. P. W. (1969), *Journal of Animal Ecology* **38**, pp.425–457, Blackwell Scientific Publications Ltd; *Figures 7.14 and 7.16:* Clutton-Brock, T. H., Guiness, F. E. & Albon, J. D. (1982), *Red Deer*, Edinburgh University Press; *Figure 7.15:* data from Palmblad, N. (1968), 'Competition studies on experimental populations of weeds with emphasis on the regulation of population size', *Ecology*, **49**, pp.26–34; *Figure 7.17:* Albon, S. D. & Clutton-Brock, T. H. (1988), 'Estimated annual heather production' in Usher, M. B. & Thompson, D. B. (1988), *Ecological Change in the Uplands*, p.99, Blackwell Scientific Publications Limited, © 1988 by The British Ecological Society; *Figures 7.18 and 7.19:* McEvedy, C. & Jones, R. (1978), *Atlas of World Population History*, Penguin Books Ltd; *Figure 7.20:* Sattaur, O. (1989), 'The shrinking gene pool', *New Scientist*, **123**, 29 July, 1989, p.41, World Press Network Ltd.

Photographs
p.14: Cumbria County Council, Kendal Library/Geoffrey Berry Archive; *p.23*: Cambridge University Collection of Air Photographs; *p.29*: Whitehaven Museum, photograph owned by Miss Combe; *p.31*: portrait of Wordsworth: Trustees of Dove Cottage; *p.39*: British Nuclear Fuels plc; *p.57*: J. Allan Cash; *p.61*: Mary Jelliffe/Ancient Art and Architecture Collection; *p.64*: Broads Authority; *p.66*: Bruce Coleman; *p.76*: Greenpeace/Morgan; *p.77: Farmers Weekly*.

Tables
Table 6.2: Steele, R. C. & Welch, R. C. (eds) (1973), *Monks Wood: A nature reserve record*, p.289, The Nature Conservancy Council; *Table 7.2: Demographic Yearbook* 1981, United Nations.

Index